ICE ISLAND

Expedition to Antarctica's Largest Iceberg

Dedicated to my wife
Austen

All proceeds from this book support ocean conservation and exploration projects.

ICE ISLAND

Expedition to Antarctica's Largest Iceberg

Gregory S. Stone

Photography: M. J. Adams

ABOUT THE AUTHOR

Dr. Stone has spent much of his twenty-five year career wet and, sometimes, very very cold. As a marine biologist, diver and explorer, Dr. Stone spends a good bit of time submerged, investigating the oceans and exploring their mysteries. When on dry land, he turns his attention to integrating marine science, conservation, technology, policy and communications.

His pioneering research in the Antarctic, Atlantic and Pacific oceans and his marine policy work are detailed in more than 60 publications, including scientific journals such as *Nature, Marine Mammal Science* and *Biological Conservation*.

As Editor of the international *Marine Technology Society Journal*, Dr. Stone is also an acknowledged leader in marine engineering and technology. He was the first non-Japanese scientist to dive in the world's deepest-diving research submarine, Japan's *Shinkai 6500*. The National Oceanographic and Atmospheric Administration's undersea saturation habitat *Aquarius* has been his home on several week-long research missions.

Currently the Vice President of Global Marine Programs at the New England Aquarium in Boston, Massachusetts, Dr. Stone is also a pioneer in marine conservation, having established the first aquarium-based conservation program. As part of the program, Dr. Stone initiated and produces an award-winning series of conservation films for aquarium, museum and zoo audiences. He also leads ocean science expeditions, writes articles for the *National Geographic Magazine*

and appears in numerous documentaries. He founded the Marine Conservation Action Fund, which provides funding for urgent projects around the world. In addition, he leads the Phoenix Islands Primal Ocean Project and is a world authority on the conservation and management of the endangered Hector's dolphin in New Zealand.

Dr. Stone has received many awards throughout the years, including the prestigious Pew Fellowship for Marine Conservation, the U.S. Navy and National Science Foundation Antarctic Service Medal, the John Knauss Marine Policy Fellowship, and the National Science Foundation/Japan Science and Technology Postdoctoral Fellowship, among others. He is also a Fellow of The Explorers Club.

Dr. Stone holds a Ph.D. in marine science, a master's degree in marine policy and a bachelor's degree in human ecology.

Although Dr. Stone has lived in New Zealand, Japan and other parts of the United States, he is a native New Englander, having first gotten his taste of diving in cold water as a teenager. When not diving in the South Pacific or studying dolphins in New Zealand, he lives in Boston with his wife and collaborator, Austen Yoshinaga.

This book is based on Dr. Stone's third expedition to Antarctica. The story was originally featured in the December 2001 issue of *National Geographic Magazine*.

EXPEDITION TEAM

Co-Leaders

Gregory S. Stone, Ph.D. (USA): Chief Scientist, Diver and *National Geographic Magazine* Writer

Wes Skiles (USA): Filmmaker, Diver and *National Geographic Magazine* Photographer

Science Team

Porter Turnbull (USA): Expedition Physician, Ornithologist and Science Diver

Carlos Olavarría (Chile): Molecular Biologist and Marine Mammal Specialist

Film Crew

Don Anderson (New Zealand): Soundman

Bob Clark (USA): Assistant Cameraman

Jill Heinerth (Canada): Television Documentary Producer, Dive Master and Exploration Diver

Paul Heinerth (Canada): Exploration Diver and Cameraman

Keith Moorehead (USA): *National Geographic Magazine* Technician and ROV Pilot

Ship's Crew

Iain Kerr (Scotland and New Zealand): Master (On leave from the New Zealand Maritime Safety Authority)

Robert Williamson (New Zealand): First Mate (On leave from Victoria University)

Matthew Jolly (New Zealand): Second Mate

Tony Campbell (New Zealand): Chief Engineer (From the Bluff Fishing Fleet)

John Spruit (New Zealand): Second Engineer

Laurie Prouting (New Zealand): Helicopter Pilot

Mike Gerzevitz (USA): Seaman

Carol Anne Chowning (USA): Ship's Cook

Nigel Jolly (New Zealand): Owner of *Braveheart*

Matthew Jolly sets
foot on the expedition's
first iceberg after a
2000-mile journey
across the Southern
Ocean.
Photography: G. Stone

TABLE OF CONTENTS

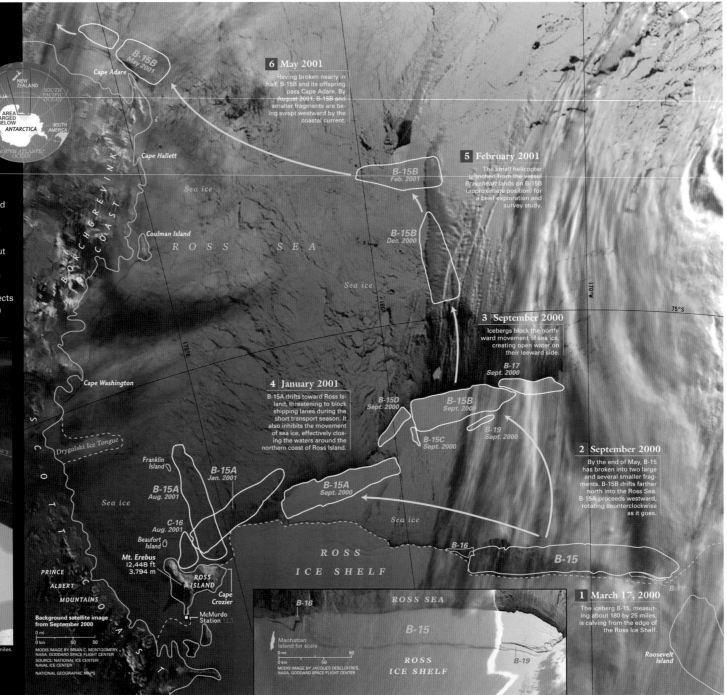

Ice Islands
Giant Bergs Adrift

On March 17, 2000, a satellite recorded a rare event at the edge of Antarctica's Ross Ice Shelf, a dynamic feature almost as big as Texas and as flat as a frozen pond. A gigantic iceberg, given the homely official name B-15, was separating from the shelf and moving offshore. How to inspect it in that frigid expanse of hostile solitude? A team of scientists and divers, a boat crew, a helicopter pilot, and others funded in part by National Geographic put out from New Zealand. In January 2001 they sailed south to study B-15, which had broken into fragments, and to assess the physical and biological effects of those and other icebergs floating in the Ross Sea.

1 March 17, 2000
The iceberg B-15, measuring about 180 by 25 miles, is calving from the edge of the Ross Ice Shelf.

2 September 2000
By the end of May, B-15 has broken into two large and several smaller fragments. B-15B drifts farther north into the Ross Sea. B-15A proceeds westward, rotating counterclockwise as it goes.

3 September 2000
Icebergs block the northward movement of sea ice, creating open water on their leeward side.

4 January 2001
B-15A drifts toward Ross Island, threatening to block shipping lanes during the short transport season. It also inhibits the movement of sea ice, effectively closing the waters around the northern coast of Ross Island.

5 February 2001
The small helicopter launched from the vessel *Braveheart* lands on B-15B (approximate position) for a brief exploration and survey study.

6 May 2001
Having broken nearly in half, B-15B and its offspring pass Cape Adare. By August 2001, B-15B and smaller fragments are being swept westward by the coastal current.

Background satellite image from September 2000

Scale varies in this perspective. Lyttelton to Cape Hallett is 2,000 miles.
IMAGE BY CARTOGRAPHIC APPLICATIONS LAB, JPL, NASA

MODIS IMAGE BY BRIAN C. MONTGOMERY, NASA, GODDARD SPACE FLIGHT CENTER
SOURCE: NATIONAL ICE CENTER
NAVAL ICE CENTER
NATIONAL GEOGRAPHIC MAPS

Manhattan Island for scale
MODIS IMAGE BY JACQUES DESCLOITRES, NASA, GODDARD SPACE FLIGHT CENTER

CHAPTER 1

The Largest Moving Object

B-15 was, at the time of its birth, the largest moving object on the earth, easily seen with the naked eye from space.

The Antarctic sun glowed red on the horizon at 1 a.m. A mile from the deck of *Braveheart* an iceberg the size of six city blocks—the iceberg to which we'd been tethered just hours before—heaved upwards, one end pausing high in the air like the bow of a foundering ship. When it crashed back down, waves swept through the waters of Cape Hallett. As the boat slapped against the bumpy sea, the iceberg rose again and the upper end seemed to explode. Shards of ice rained down, covering two square miles like shattered crystal.

We were watching a part of our world come apart. Expedition co-leader Wes Skiles and I were below decks eating a meal when someone yelled to come up on deck... quickly! We raced through the companionways, up the stairs, bumping our heads and grabbing cameras on the way. Almost too stunned to take pictures, we watched, mute, as the giant piece of ice vanished.

We were possibly the only people to record an iceberg exploding at such close range. As we stood on deck watching the seemingly solid mountain of ice disintegrate before our eyes, we were reminded of the very real dangers surrounding us in the world of ice we had come to study.

Our ultimate quarry was cruising the Ross Sea some 125 miles from Cape Hallett. Its name was B-15B, a 1,900 square mile chunk of the original B-15 which, when it calved from the Ross Ice Shelf in March 2000, had an upper surface area of about 4,500 square miles (170 miles by 26 miles) and was nearly half a mile (2,200 feet) thick. B-15 was comparable in size to Connecticut or Jamaica, and contained enough fresh water, in the form of a thousand cubic miles of ice, to supply the United States for five years.

Geophysicist Doug MacAyeal from the University of Chicago tracked B-15 from satellite images and nicknamed it "Godzilla." After calving, B-15 drifted west, bumping and smashing along the edge of the Ross Ice Shelf and eventually breaking up into smaller, but still gigantic, pieces named B-15A, B-15B and B-15C, and so on.

A berg with the dimensions of B-15 comes along once or maybe twice in a lifetime. The U.S. National Ice Center in Suitland, Maryland, which has been tracking Antarctic icebergs for 25 years, had never recorded a berg as big. For nine years, satellite images recorded cracks in the Ross Ice Shelf as they spread and set B-15 loose, making it the largest "fast" moving object on the planet and sending it on a journey that would eventually lead to us.

A month after B-15 was born, photographer and filmmaker Wes Skiles and I decided to launch an expedition to study and explore B-15 and other large icebergs around Antarctica.

On the day Wes called me, I was diving in Bermuda on a coral reef project. Under a tropical sun, it was easy for me to jump at this ambitious proposal but later, sliding across the ice-covered deck and wedged into my bunk in the towering seas, I would question my sanity. But it had been 13 years since

I had dived into Antarctica's clear, frigid waters, and I had often dreamed of returning. I could not wait to get back to the most exhilarating diving I had ever done.

It took a few hundred phone calls, numerous meetings at the National Geographic Society, a winter training session in the Colorado Rockies, and ambitious fund-raising efforts, but we were able to pull the expedition together in just seven months.

We were especially thrilled to have backing from National Geographic, for which Wes filmed and photographed the expedition and I wrote an article (December 2001).

The mission of the expedition was scientific. As a marine biologist, I was interested in the ecology of large icebergs, which are really floating ice-islands. There is little information on how large icebergs affect the ocean and the distribution of animals around them. And, since ice is melting globally in more places and at higher rates than ever recorded, I wanted to dive and sample these melting giants. Any information we gleaned about these icebergs would help us understand the effects of global warming.

We decided to head for B-15B, the second-largest chunk of B-15 and the one farthest north in the Ross Sea. At the time, it measured approximately 60 miles by 25 miles.

Top: Wes Skiles, expedition co-leader, *National Geographic Magazine* photographer, diver and filmmaker.
Photography: G. Stone

Bottom: *Braveheart* started life as the Japanese fisheries patrol vessel *Genkai*. She is 118 feet in length with an 882 kW Niigata engine and variable pitch propeller. She was not registered as an ice-strengthened vessel, but she deserves that designation from her time on the *Ice Island* expedition in the heavy ice of the Ross Sea.
Photography: P. Turnbull

The *Ice Island* expedition set out in search of the world's largest iceberg, B-15, in the Ross Sea. Here, Antarctic mountains rise above the waters and icebergs of the Ross Sea near Cape Adare.
Photography: G. Stone

3

Aurora comes to the rescue of the 1917 Imperial TransAntarctic Expedition. This vessel carried men and supplies to lay food depots for Sir Ernest Shackleton's planned crossing of the Antarctic continent on foot. These vital supply caches were to be found by Shackleton and his men at critical places along a route from the Weddell Sea on the Ross Ice Shelf. However, Shackleton never set foot on the Antarctic continent during the expedition. His vessel, the *Endurance,* was crushed by the ice and sank.
Photography: Courtesy of the Joyce Collection, Canterbury Museum.

4

CHAPTER 2

Those Who Went Before

What motivated these men, how did they get there, what happened to them, what could I learn from them and what might we expect on our journey?

Reading the accounts of their expeditions late into the night, my eyes gritty with sleep, I was filled with admiration for these irrepressible explorers, their hardiness, determination and resourcefulness. At other times, reading these same accounts filled me with apprehension about our voyage.

Once I had decided to co-lead an Antarctic expedition, I became passionately interested in the early explorers. The smallest details of their journeys fascinated me, as I planned a trip to one of the most desolate places on earth — hoping to avoid some of their mistakes. Our trip would have more in common with the early explorers of the Ross Sea than any contemporary "Antarcticans." Among these early explorers were Sir James Clark Ross, Captain Robert Falcon Scott, Sir Ernest Shackleton, Captain Roald Amundsen and Sir Douglas Mawson.

Aboard ships with names like the *Erebus, Terror, Discovery, Fram, Nimrod, Terra Nova* and *Aurora*, these men discovered and lived in the region. What motivated these men, how did they get there, what happened to them, what could I learn from them and what might we expect on our journey? Reading the accounts of their expeditions late into the night, my eyes gritty with sleep, I was filled with admiration for these irrepressible explorers, their hardiness, determination and resourcefulness. At other times, reading these same accounts filled me with apprehension about our voyage.

My previous trips to the Antarctic had been to the other side of Antarctica, the Peninsula region, a much easier place to get to and work. People call it the "banana belt" of Antarctica for its relatively mild summer weather. But we were going to the Ross Sea, a completely different kind of trip: colder, a 14-day voyage, gale-force winds, much higher risk overall, but closer to the heart of Antarctica in every way. More than 12,000 people travel to Antarctica each year, but most either go in airplanes to modern research bases or in larger ships to a few tourist areas in the relative comfort of the Antarctic Peninsula's

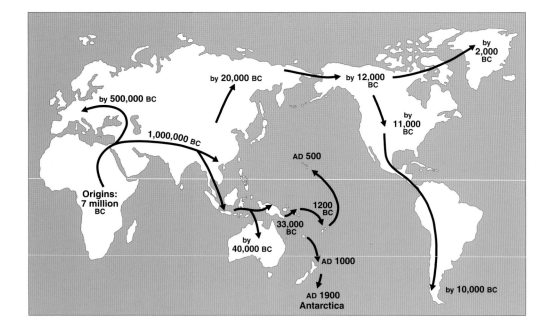

The Long, Slow March of Exploration

Exploring our world is part of human nature. From our DNA, our biology and the fossil record, it is apparent that humans originated in Africa, and from there proceeded to populate every major region of the planet. The first destinations were Europe and Asia some 500,000 to 1,000,000 years ago across the land bridge that is now the Middle East. This distribution remained fairly stable until the next great expansion began 50,000 years ago, when humans walked, with the help of a lower sea level, through what is now Southeast Asia, Indonesia and Australia. Then, between 14,000 and 20,000 years ago when sea levels were still low, humans wandered across the Bering Sea land bridge and into the wilderness that was to become North, Central and South America.

Everywhere these early people went they encountered pristine forests, plains and untouched populations of animals. These conditions changed relatively quickly with their arrival, and now there are only remnants of these natural ecosystems left.

These prehistoric explorers lived through the periodic advance of the great ice sheets. During the most recent ice age, humans followed the retreating glaciers and settled the northern Arctic regions, thereby establishing themselves throughout the Northern Hemisphere.

Coincident with the last ice age, humans began the slow march to the south through the Americas, advancing about 100 miles per generation, from North America to the tip of South America. Once we reached Patagonia, our final major expansion was complete, making us the most widely distributed mammal on the planet. What remained untouched at this stage were New Zealand, the South Pacific Islands and Antarctica.

Intrepid Polynesian, Melanesian and Micronesian mariners in canoes established colonies on the South Pacific islands and New Zealand between 1,000 and 20,000 years ago, but the gigantic ice-covered land of Antarctica remained completely out of reach of the small boats and canoes of the day. Antarctica slept its cold sleep and awaited a new wave of explorers who had the same enduring human spirit of curiosity and exploration, but better equipment.

Top: Approximate routes and dates for human migrations and settlements across the continents. Although Antarctica is the world's fifth largest continent, it was not settled until the 1900s. *Adapted from **Guns, Germs and Steel**, J. Diamond, with permission from W. W. Norton & Co., 1996.*

summer climate. Our expedition would take a small ship and go to the more southerly, rarely-visited offshore waters of the Ross Sea.

We were going to explore parts of Antarctica, the fifth largest continent (about twice the size of Australia), which covers 10 percent of the world's total land area. It contains most of the fresh water on the planet in the form of an ice sheet, which is really a series of massive glaciers that cover 99.5 percent of the land area and tower more than two miles above the continent's soil. B-15 was originally part of this ice sheet. This tremendous ice sheet began to form more than 10 million years ago, but because most of the ice continually flows outward and off the continent into the ocean, the ice there today is usually no more than 100,000 years old. This ancient icy world makes the continent what some call a Pleistocene relic. Winters are six months of nearly complete darkness, with temperatures and winds fiercer than any other place on earth.

Though there are no land mammals in Antarctica, there are more marine mammals — seals and whales — than anywhere else. Antarctica is home to 33 million seals and more than 500,000 whales, but the diversity is not high: only six seal species and nine whale species inhabit its waters, including the largest animal to ever live, the blue whale (*Balaenoptera musculus*). For sheer numbers and size of individuals, Antarctica is a place of mammals.

6

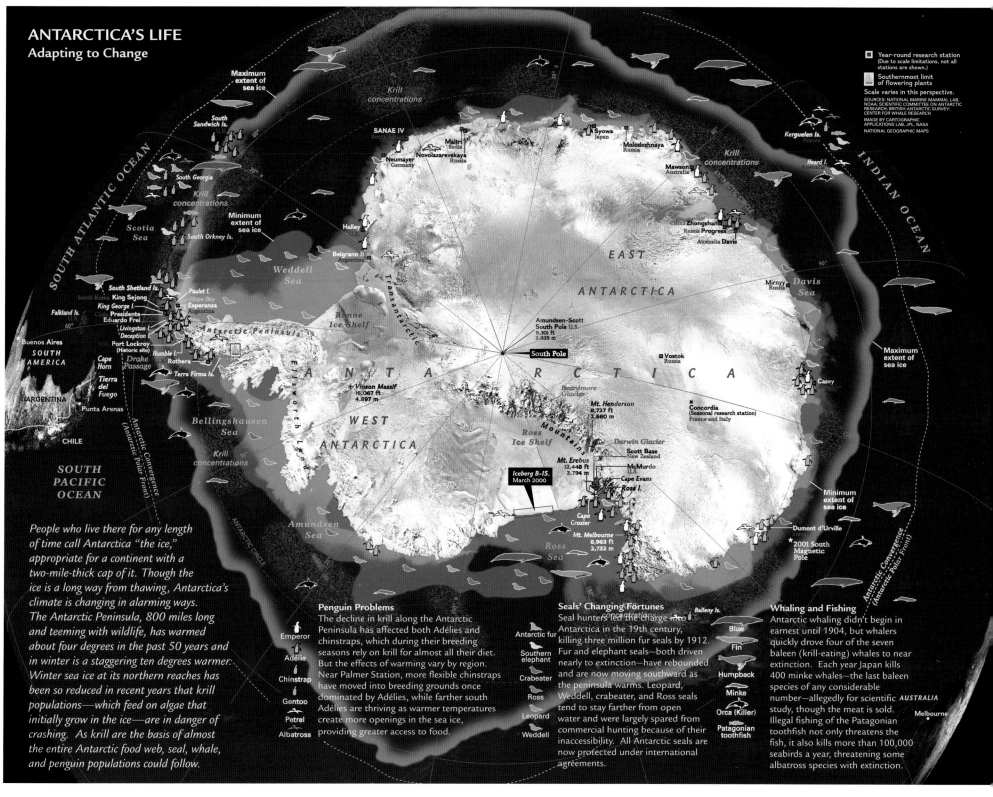

ANTARCTICA'S LIFE
Adapting to Change

Year-round research station
(Due to scale limitations, not all stations are shown.)

Southernmost limit of flowering plants

Scale varies in this perspective.

SOURCES: NATIONAL MARINE MAMMAL LAB, NOAA; SCIENTIFIC COMMITTEE ON ANTARCTIC RESEARCH; BRITISH ANTARCTIC SURVEY; CENTER FOR WHALE RESEARCH

IMAGE BY CARTOGRAPHIC APPLICATIONS LAB, JPL, NASA

NATIONAL GEOGRAPHIC MAPS

People who live there for any length of time call Antarctica "the ice," appropriate for a continent with a two-mile-thick cap of it. Though the ice is a long way from thawing, Antarctica's climate is changing in alarming ways. The Antarctic Peninsula, 800 miles long and teeming with wildlife, has warmed about four degrees in the past 50 years and in winter is a staggering ten degrees warmer. Winter sea ice at its northern reaches has been so reduced in recent years that krill populations—which feed on algae that initially grow in the ice—are in danger of crashing. As krill are the basis of almost the entire Antarctic food web, seal, whale, and penguin populations could follow.

The text, written by R. Smith, and map are reprinted with permission from *National Geographic Magazine.*

Emperor
Adélie
Chinstrap
Gentoo
Petrel
Albatross

Antarctic fur
Southern elephant
Crabeater
Ross
Leopard
Weddell

Blue
Fin
Humpback
Minke
Orca (Killer)
Patagonian toothfish

Penguin Problems

The decline in krill along the Antarctic Peninsula has affected both Adélies and chinstraps, which during their breeding seasons rely on krill for almost all their diet. But the effects of warming vary by region. Near Palmer Station, more flexible chinstraps have moved into breeding grounds once dominated by Adélies, while farther south Adélies are thriving as warmer temperatures create more openings in the sea ice, providing greater access to food.

Seals' Changing Fortunes

Seal hunters led the charge into Antarctica in the 19th century, killing three million fur seals by 1912. Fur and elephant seals—both driven nearly to extinction—have rebounded and are now moving southward as the peninsula warms. Leopard, Weddell, crabeater, and Ross seals tend to stay farther from open water and were largely spared from commercial hunting because of their inaccessibility. All Antarctic seals are now protected under international agreements.

Whaling and Fishing

Antarctic whaling didn't begin in earnest until 1904, but whalers quickly drove four of the seven baleen (krill-eating) whales to near extinction. Each year Japan kills 400 minke whales—the last baleen species of any considerable number—allegedly for scientific study, though the meat is sold. Illegal fishing of the Patagonian toothfish not only threatens the fish, it also kills more than 100,000 seabirds a year, threatening some albatross species with extinction.

FINDING ANTARCTICA

The Antarctic continent was not even
known to exist with certainty until several
explorers in 1820 made sightings of the
Antarctic Peninsula, confirming that
there was land to the south. One of these
explorers was an American sealing captain,
Nathaniel Palmer of the *Hero*, who made this entry in his log
on November 17, 1820:

> *"These 24 hours commences with fresh breezes from
> Swest and pleasant at 8 P.M. got over under the Land found sea
> filled with immense Ice Bergs – at 12 hove Too under the Jib
> Laid off & on until morning – at 4 A.M. made sail in shore and
> discovered a strait – Tending SSW & NNE – it was Literally filled
> with Ice and the shore inaccessible we thought it not Prudent to
> Venture in ice – Bore away to the Northern & saw 2 small islands
> and the shore every where Perpendicular we stood across toward
> friesland Course NNW the Latitude of the mouth of the strait
> was 63-45S." [sic]*

Captain Nathaniel
Palmer, the 21-year-old
captain of the *Hero*,
may have been the first
to record a sighting of
Antarctica in 1820.
*Photograph courtesy of
Canterbury Museum.*

Captain Palmer was a mere 21 years old when he
captained a vessel exploring some of the roughest, coldest
and most remote waters in the world. That he had the
maturity, experience and skills to command a group of
men and safely sail to Antarctica and back at that age
astonishes me.

Because of its relative proximity to the southern tip
of South America and its tolerable climate, the Antarctic
Peninsula was the first part of Antarctica discovered and
explored. To live year-round in the Antarctic Peninsula, you
still need equipment that looks more suited to a space
mission, but it is warmer than other areas in Antarctica by
some 10-20 degrees Fahrenheit during summer months.

8

Pack ice has always been a problem for Antarctic explorers. On our expedition, pack ice often blocked *Braveheart's* way in the Ross Sea. While waiting for the ice to break up, we occasionally took walks on the ice.
Photographer: C. Olavarría

Captain James Clark Ross, British Naval Expedition, 1839-43. Captain Ross discovered the North Magnetic Pole in 1831. Eight years later he failed in his attempt to find the South Magnetic Pole but he made perhaps more important discoveries, finding and naming the Ross Sea and Ross Ice Shelf.
Photograph courtesy of Canterbury Museum.

THE OTHER SIDE: ROSS DISCOVERS THE ROSS SEA

On the other side of Antarctica is the Ross Sea, the home of B-15. Though colder, icier and windier than the Peninsula region, the Ross Sea became the staging area for early attempts to reach the South Pole because it was the closest access point by sea to the very heart of Antarctica.

About 20 years after Captain Palmer sighted the Antarctic Peninsula, a British explorer, Sir James Clark Ross, commanded the 105-foot *Erebus* and the 102-foot *Terror* during the Antarctic expedition of 1839-43. Ross's discovery of the North Magnetic Pole in the Arctic at the age of 31 inspired him to seek the South Magnetic Pole eight years later. Although Ross never found the South Magnetic Pole, a prize that awaited Sir Douglas Mawson some 70 years later, he signposted and blazed the trail deeper into the Antarctic than anyone had done before, making numerous geographically important discoveries along the way. He delineated the Ross Sea and the Ross Ice Shelf, both of which now bear his name.

The Ross Sea is covered in sea ice for most of the year. There is a brief window in January and February when it breaks up enough that a ship can poke its way through the ice, but it never completely clears out. It was during this window that Captain Ross entered the Ross Sea, and it was this same window that we used to enter the Ross Sea 162 years later aboard *Braveheart*.

The ice-covered entrance to the Ross Sea is daunting even to today's modern ships. Ross's decision to sail his wooden, square-rigged ships into the ice, with no knowledge of what lay beyond, embodies the spirit and determination that defined these early explorers. We will never know what went on in

Ross's mind as he looked into that ice, knowing there was a reasonably good chance he and his men would not come out.

Ross is best known for his discovery of the Ross Ice Shelf. Ice shelves are massive glacial sheets that slide off the continent and float out over the ocean while remaining attached to the glacier on land. They were considered a "natural wonder" of the world at the time and had never been seen until the discovery of the Ross Ice Shelf. Here is how Ross described it in his ship's log entry dated January 28, 1841:

"As we approached the land under all studding sail, we perceived a low white line extending from its eastern extreme point as far as the eye could discern to the eastward. It presented an extraordinary appearance, gradually increasing in height, as we got nearer to it, and proving at length to be a perpendicular cliff of ice, between one hundred and fifty and two hundred feet above the level of the sea, perfectly flat and level at the top, and without any fissures or promontories on its even seaward face...What was beyond we could not imagine..."

At the time he gazed at the seaward margin of the Ross Ice Shelf and wrote these words, the mass of ice that became B-15 was somewhere well back in the slow stream of moving ice. After Ross discovered and explored the region, the world forgot about it for half a century.

Top: Captain Robert Falcon Scott, Royal Navy. Scott is best known for his unsuccessful attempt to be the first to reach the South Pole, a race he lost to Captain Roald Amundsen in 1911. Scott and his crew of three died on their return trek across the ice.
Photograph courtesy of Canterbury Museum.

Bottom: Scott's first boat, *Discovery*, leaving Lyttelton Harbour, British Antarctic Expedition, 1901-04. Shackleton was also on this trip, and sledged with Scott and biologist Edward Wilson to 82 degrees south, farther south than anyone had ever been.
Photograph courtesy of Kinsey Collection, Canterbury Museum.

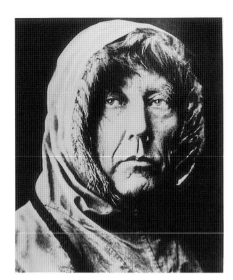

The Next Generation: The Race for the South Pole

Interest in Antarctica surged again in 1901, when the race to find the South Pole began. The successes, failures and deaths of Captain Roald Amundsen, Sir Ernest Shackleton and Captain Robert Falcon Scott are well

accounted in many works. The race ended when Norwegian Amundsen beat British Scott to the South Pole by a little more than a month in December 1911. This was a heart-wrenching defeat for Scott who then died on his over-land journey from the South Pole back to his ship, the *Terra Nova*.

Despite never achieving any of his expedition goals, Shackleton is perhaps the most famous of the Antarctic explorers; He was an extraordinary leader with the presence of mind to make the right decisions, even in the face of starvation and death. On his last expedition, aboard the *Endurance,* he managed to save some of the now-famous photographic plates that give us a rare glimpse into this heroic age of Antarctic exploration. In recent years, a number of accounts of the incredible disasters and triumphs of his expeditions have been published, and they are well worth reading (see page 72).

Individual successes and failures aside, the most significant contribution of these early explorers was the establishment of year-round accommodations for people on the continent. Reaching Antarctica marked the completion of the last major expansion by humans, the most mobile and adaptable land-dwelling vertebrate species to have ever lived.

During this phase of exploration, both Scott and Shackleton used the New Zealand port of Lyttelton as their point of departure because it was the closest to the Ross Sea. Following in their footsteps, we decided to depart from Lyttelton as well.

Left: Antarctic explorers Sir Douglas Mawson (left) and Admiral Richard Byrd, mid-1930s. *Photograph courtesy of S.P. Andrew (Wellington), Canterbury Museum.*

Right: With Captain Scott in command, the *Terra Nova* leaves Lyttelton Harbour, November 26, 1910, British Antarctic Expedition, 1910-13. It was on this expedition that Scott, beaten to the South Pole by Amundsen, died on his overland journey back to his ship. *Photograph courtesy of Kinsey Collection, Canterbury Museum.*

13

Top: Aboard *Braveheart*, expedition team members take one last look at Lyttelton before heading off to the Ross Sea.
Photography: J. Heinerth

Right: January 17, 2001 dawned warm and sunny as Wes Skiles (center) and the rest of the expedition team finalized preparations for *Braveheart's* departure.
Photography: G. Stone

Getting There: Shipboard Life

We were in the only place on the planet where there is no land mass to impede the wind.

The Southern Ocean surrounding Antarctica covers some 13.5 million square miles, making it the fourth largest ocean. The region is notorious for having the worst sea conditions in the world, with waves 100 feet high and winds more than 100 miles per hour.

To study large icebergs and to reach B-15B we followed the lead of the early twentieth century Antarctic expeditions and enlisted both explorers and scientists. Wes and I recruited an 18-member team of divers, environmental scientists and a rough and ready New Zealand boat and crew with a go-any-where/do-anything attitude. On a sunny, warm January 17, 2001, the whole expedition team was on the docks and on deck getting everything ready to go.

The dive team included Jill and Paul Heinerth, accomplished cave divers and experts at using closed-circuit rebreathers that allow them to explore at depths and in places too dangerous for open-circuit scuba gear. They would dive beneath and inside icebergs, including the one that disintegrated before our eyes.

Porter Turnbull, the ship's medical officer and ornithologist, was my ice diving partner. Chilean scientist Carlos Olavarría was our molecular biologist and marine mammal specialist. As the overall chief scientist and expedition co-leader, I would focus my research on the marine life around, inside and under large icebergs. Together, Porter, Carlos and I were the science team.

Braveheart's crew was mostly from New Zealand, another similarity to expeditions of the early twentieth century. Scott and Shackleton recruited crew members from New Zealand ports prior to departure. There is something extraordinary in New Zealanders that often makes them excellent expedition members and shipmates. The people who sign on to these journeys crave adventure and are willing to take risks and do any job that needs to be done, whether or not it is in their job descriptions. Our crew was a lively, fun bunch with a knack for storytelling and a love of jokes of all kinds.

Antarctic explorers often used the New Zealand port of Lyttelton as their point of departure because of its proximity to the Ross Sea. Following in the footsteps of explorers like Scott and Shackleton, we left from Lyttelton Harbour on January 17, 2001.
Photography: J. Heinerth

When I would go to the galley late at night for a cup of tea, I would usually find second engineer John Spruit awake and reading one of his many motorcycle magazines. John wore a sweater with about 50 holes in it. Looking up over the edge of his magazine with a wild look in his eyes, he would offer me jokes that I could barely understand through his strong New Zealand accent. John and Matt Jolly, the second mate, were in charge of laundry. They would wash our clothes each evening in a washing machine and then hang the wet garments below decks all over the running diesel engine, which put out plenty of heat to dry them. Clothes dried in a warm engine room took on a distinct diesel scent, one of the many joys of expedition life.

Our helicopter owner and pilot, Laurie Prouting, was a sheep farmer from New Zealand's South Island. Laurie's farm was named Mesopotamia, perhaps in tribute to its large size, and he needed his own helicopter to get around it. He had never been to the Antarctic before, but when he heard about the trip, he signed on and flew his tiny Hughes helicopter to the dock, landed on *Braveheart's* aft deck, removed the blades, covered her in a tarp and strapped her down for the long, wet journey across the Southern Ocean. This simple act of flying from a sheep farm and heading to Antarctica really exemplifies the New Zealand spirit. Laurie's helicopter and excellent flying skills played a key role in helping us find our way through the pack ice of the Ross Sea and in finally reaching B-15B.

Paying Homage to History: Quail Island

In the warm New Zealand summer sun, we were loading equipment and supplies, stowing provisions, tying things down and doing all the important last-minute tasks. With no stores or refueling stations between Lyttelton and Antarctica, we needed to bring absolutely everything we would use for the eight-week journey. The list started with more than 100,000 liters of diesel to feed the 882 kW Niigata engine that was to churn non-stop for the next 60 days. We also loaded several tons of food to feed the 18 expedition members, dive gear, camera gear, scientific gear and thousands of other items essential to our exploration and survival. We didn't take water because the ship had a desalination system that made fresh water from sea water, but, as we found out, the fresh water was still a little salty and peculiar-tasting.

Left and center: Laurie Prouting, a New Zealand sheep farmer, took a holiday from his sheep to do a spot of helicopter piloting in the Antarctic.
Photography: left, G. Stone, center, C. Olavarría

Right bottom: In order to bring the more than 100,000 liters of fuel we needed, we had to store extra tanks on *Braveheart's* deck.
Photography: G. Stone

Possibly to keep me occupied during the final preparations, someone suggested that I visit Quail Island. Quail Island has a history of hosting Antarctic explorers. Originally a leper colony, the small island is less than a mile from the New Zealand port of Lyttelton, but enough harbor water separates it from the township to create a biological barrier.

By the early twentieth century, when leprosy was no longer an issue, it was used as a quarantine island for incoming farm livestock for this region of New Zealand. Because Scott and Shackleton brought dogs and ponies from Europe for their Antarctic explorations, they had to kennel and stable them on the island before leaving for the Ross Sea to prevent any diseases from infecting local New Zealand animals.

We launched one of our skiffs and motored across to Quail Island to see what was left. All that remained of these early expeditions were remnants of the pony stables and one restored kennel. Standing in the spot where Shackleton and Scott may have stood long ago, I was inspired and ready to be off on an adventure of my own.

While I was off reliving history, the *Braveheart's* crew had finished loading, stowing and tying down, and we were ready to go.

Top left: Captain Scott with ponies on Quail Island, the historic quarantine site for expedition dogs and ponies, November 1910, British Antarctic Expedition, 1910-13 *(Terra Nova)*. *Photograph courtesy of Hutton Collection, Canterbury Museum.*

Top right: Captain and Mrs. Scott inspecting the expedition dogs, Quail Island, November 1910, British Antarctic Expedition, 1910-13 *(Terra Nova)*. *Photograph courtesy of O'Neill Collection, Canterbury Museum.*

Bottom: Quail Island: A restored dog kennel from Scott's expedition draws tourists interested in the Antarctic expeditions. *Photography: G. Stone*

Onward Ho!

After returning from Quail Island, I found myself standing in the warm summer sun preparing to board *Braveheart* from wharf number four in Lyttelton Harbour.

Braveheart's 118 feet suddenly looked very small to me. She was a former Japanese research vessel with 2,000 miles of the roughest seas in the world ahead of her. Austen, my wife, looked worried as she came to see me off and asked, "Where's the mother ship?" *Braveheart* seemed very small to her, too. I tried to convince her that many people had traveled to the Antarctic in boats this size, but I could tell it wasn't working. As I stood there with her, I thought about the stories I had heard of boats capsizing in the rough seas south of New Zealand and shared her dismay. I was taking a small boat into rough, frozen waters and then diving near monster icebergs. Had I lost my mind? But there was no turning back now. I had put too much work into this project. Austen kissed me good-bye (with a prayer). I boarded *Braveheart*, propped my elbows on the rails and got ready to also say good-bye to dry land.

A crowd of friends, family and local people crowded around *Braveheart* as we cast off dock lines and left the port. The steep green hills of Lyttelton surrounded our tiny ship, which was dwarfed by the larger freighters, fishers and cruise ships in port at the time. Good-bye warmth, good-bye privacy, good-bye comforts. It was January 17, 2001, and we were on our way. Our escort leaving Lyttelton was a group of Hector's dolphins (*Cephalorhynchus hectori*) that swam, as dolphins love to do, in our bow wave.

We were soon in the rolling sea of the Southern Ocean. As we moved farther and farther from land, the water turned darker and bluer, colors that I loved and celebrated. I was at sea once again, surrounded by ocean waves, soaring albatross and wind. I got my sea legs quickly, but several other expedition members were not so lucky. Even those of us not actually sick in bed had a hard time eating and we felt queasy most of the time. This went on for weeks.

Porter and I were on deck surveying birds and marine mammals whenever possible, when weather and waves would not wash us overboard. In the first day we saw five species of albatross, a sperm whale (*Physeter macrocephalus*) and more dolphins. The farther south we moved the bigger the swells grew and the more *Braveheart* pitched and rolled. Soon it was too rough to be on deck counting birds and it was so rough that walking anywhere on the ship was a challenge; we banged our heads into walls, stumbled like drunks down companionways and spilled whatever drink and food we could handle onto the table, the floor and ourselves. Those of us who were not steering the ship, tending the engine or performing other essential duties were wedged in our bunks, trying to read and wondering what lay ahead.

Quail Island, Lyttelton Harbour, New Zealand, circa 1953. Enough water separates Quail Island from Lyttelton to create a biological barrier. This is where Scott and Shackleton quarantined their expedition dogs and ponies prior to departure.
Photograph courtesy of V. C. Browne, Harrow Collection, Canterbury Museum.

Facing page inset: At sea at last, an albatross soars above *Braveheart*.
Photography: P. Turnbull

Facing page: Hector's dolphins, beautiful but endangered, live only in New Zealand and are common in the waters around Lyttelton. Four mothers with calves escorted *Braveheart* as we left Banks Peninsula for the Ross Sea.
Photography: G. Stone

Top: Porter Turnbull recorded sightings of all the oceanic seabirds we saw during the expedition.
Photography: G. Stone

Bottom: Porter and Nigel Jolly, *Braveheart's* owner, had the difficult task of securing fuel tanks that had come loose while *Braveheart* swayed in a swell.
Photography: G. Stone

Every expedition has several defining moments, unexpected events that become your most accessible memories of the trip, those most buoyant images and feelings that often float to the surface of your mind. These are the experiences you tell people when they ask you about the expedition. The first such moment happened on the third day out of New Zealand, not far from Campbell Island.

I was below decks in my bunk when suddenly there was a sound like an explosion, after which the ship's entire hull shuddered and gave off a deep rumble. I was thrown out of my bunk onto the floor, with clothes, books, cameras and crates bashing about in my cabin. In the galley, dishes shattered as they hit the floor. On deck, two of the massive containers of extra fuel came loose and were sliding dangerously around, threatening to crush crew members.

We had hit a big wave, probably 60 or 70 feet, and *Braveheart* was now rolled over on her side. I was on the floor, covered in fallen clothes and equipment, waiting for her to come upright. Although it was only a few seconds, it seemed much longer.

When a ship rolls far over, she has to make a decision: Will she come back and right herself or will she roll over and capsize? If another big sea hits on the same side while she is rolled over, that will push her further and may cause her to capsize. As a mariner you want your ship to snap back relatively quickly when she rolls. The snapping point is called her "righting moment." A fast righting moment is safe, but very uncomfortable because the ship jerks sharply. For comfort, passengers like slow and easy motions. The bigger the ship, the slower the motion. But when any ship gets in real trouble, and by that I mean really rough seas with a breached hull, its motion (rolling and pitching) slows considerably as the ship takes on water. The lower the vessel is in the ocean, the more stable it becomes and the less quickly it can right itself. It is a cruel deception that when the ship is getting ready to sink, the passengers have a sense that everything will be okay because the motion slows down.

Braveheart did right herself relatively quickly, forging my confidence that she was the vessel for our trip. I struggled to the bridge in the still-heaving seas to find out how far we rolled.

There was an inclinometer on the wall. An inclinometer is a small pendulum that records the degree of swing to port (left) or starboard (right) a ship rolls. The maximum roll an inclinometer can measure is 45 degrees — and that was where ours was stuck, which meant we had actually rolled farther than that. How much farther we will never know. After that, we were more careful about slowing down for big waves.

For the next two days we pounded and crashed through seas that rose higher than *Braveheart's* bridge, but because we were going slowly, we never repeated our earlier excitement. Looking up at 60-foot walls of water topped by frothy peaks became routine. I felt diminished by these waves, as if I had moved into a larger world. Going on deck became a life-threatening activity as seas washed over the bow and down the ship's sides with incredible force.

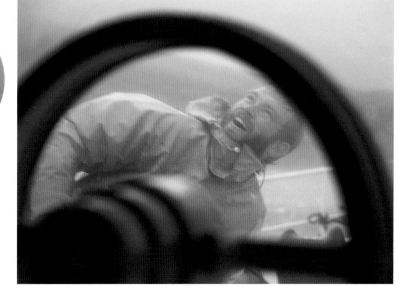

We were now deep into the Southern Ocean where the wind blows from west to east right around the globe. This is the only place on the planet without a land mass to impede the wind that builds waves seemingly forever, creating a banging, slamming, knee-bumping, stomach-wrenching, bone-drenching experience. Every day the ocean got rougher, the air colder and the hours of daylight longer.

Top: On our way to Antarctica and in the Ross Sea, high winds and big waves were always a worry. The wind blew gale force or stronger on most days. Here, a gale blows off one of the many giant tabular bergs that were originally part of B-15. *Photography: W. Skiles*

Inset: Captain Iain Kerr. *Photography: G. Stone*

Bottom right: Matt Jolly works to reset *Braveheart's* anchor after she lost her holding in the 100 mile per hour wind near Campbell Island. *Photography: G. Stone*

"An absolutely nice day." Wes recalls a moment when he caught *Braveheart's* bow crashing into a wall of wave. Winds often blew at gale force. The ship once rolled more than 45 degrees but came back up — as did, now and again, a lot of the crew's meals.
Photography: W. Skiles

THE MIDDLE OF NOWHERE

As we neared Campbell Island the wind speed rose even higher. Our wind gauge registered 100 miles per hour, hurricane force, and we were only one-third of the way to Antarctica. By going out on deck in this wind, we risked not only getting washed overboard, but also getting blown overboard. It was still impossible to sit and eat normal meals. To eat, I would go into the galley and grab whatever food was around, cradle a bowl or plate under my arm, jam my body in a passageway and stuff morsels into my mouth.

The ocean and the air were much colder now. We were taking out our winter clothes, when just a few days before we had been in shorts and t-shirts on the dock in New Zealand. It amazed me how quickly it can get cold with only a few days of southerly travel on a boat going eight knots (a little over nine miles an hour). The warm New Zealand summer was now behind us. Ahead was Campbell Island.

First discovered by Captain Frederick Hasselburg in 1810, Campbell Island is an uninhabited sub-Antarctic island. It is in the zone between temperate climates and the perpetual ice of the Antarctic. It is a way station, a place of nature and not civilization, with its own unique animals and low bushes and grasses. Campbell Island is a place that still

belongs to the birds and sea lions that nest and rest on its hills and shores.

The island rises like an oasis from nowhere in the immense Southern Ocean and has a deep, long harbor that we gratefully entered. We anchored to repack the ship, leave some of the used fuel tanks onshore (which we picked up on our return to New Zealand), and consider our next step. The stop allowed several expedition members to recover enough to eat a little.

I loved Campbell Island for its prehistoric beauty and feeling, and for its protected harbor that gave us refuge. At night, groups of endangered Hooker sea lions (*Phocarctos hookeri*) cavorted in the water around us, circling with barks and yelps.

Watching the sea lions cheered us all. After the rough seas, it was restful to enjoy their companionship and the gentle boat movement in a calm harbor.

While we were anchored at Campbell Island, groups of endangered Hooker sea lions *(Phocarctos hookeri)* cavorted in the water around us, circling with barks and yelps. These young males play fight on shore in preparation for more serious bouts as adults.
Photography: P. Turnbull

Facing page: An uninhabited sub-Antartic island, Campell Island is a way station between temperate climate and perpetual ice. A place of nature and not civilization, it has its own unique animals and plants, consisting mostly of low bushes and grasses.
Photography: P. Turnbull

On our passage from Lyttelton to Campbell Island, I learned a lot about our expedition team. Many had been visibly shaken by the rough seas and high winds, and all of us were wondering what lay ahead. We still had to cross more than a thousand miles of Southern Ocean.

There was no turning back now. You might just as well ask to get off a spaceship after a launch. I realized just how worried people were when I heard a rumor that we didn't have enough food aboard the ship for the rest of the journey. Visions of hunting and eating penguins swam in the minds of several crew members, as they remembered the grainy black and white pictures from Shackleton's expeditions. No one relished the thought of being 2,000 miles from home, freezing and hungry.

To put everyone's mind at rest, we spent a day unpacking all the stored food and laying it out on deck for inspection. We had sides of beef, legs of lamb, bags of potatoes, vegetables, fruits, canned goods and more covering the aft deck. Everyone dutifully worked through the smorgasbord, stepping up high over the bundles and intently inspecting the cache. It turned out that the rumor was started by someone who desperately wanted to turn back after the rough seas, but we were committed and after assuring everyone that there were plenty of provisions, we carried on.

The passage from Campbell Island to the ice was uneventful except for trying to eat while being tossed around,

trying to sleep and trying to read as books jumped up and down in front of our faces. During the passage from Lyttelton to Campbell, we learned the full capability of the ship and gained enormous confidence in her. We also got to know our shipmates and began to navigate the social map of our tiny vessel, which was now our whole world. Shackleton is famous for hiring an expedition member solely for his singing ability, and as we chugged along far away from those we loved, I understood his reasoning.

Our daily sense of community came at mealtime. Meals consisted of cold cereal for most breakfasts, soups for lunch, and then the standard New Zealand fare of meat (lots of lamb) and starches for dinner. We soon learned who among us had bad tempers and moods that would infiltrate the entire group at these times. We grew to know each other so well that one comment or a special inflection in a voice would tell us how people were doing. Isolation can bring out the best and worst in people. As expedition leaders, Wes and I would find a private location on deck or in our cabins to discuss personnel issues each day. Who needed more attention? How to keep people busy? The issues and needs of 18 people on a small boat were surprisingly endless.

Top: With each passing day, the air grew colder and the days grew longer. First mate Robert Williamson (left) helped Carlos Olavarría and me with marine mammal surveys when the seas weren't too rough.
Photography: P. Heinerth

Bottom: Despite the tight quarters and full schedules, Porter, Jill (above) and Paul Heinerth (below) found time and space to socialize below decks.
Photography: G. Stone

Being so far from home, any contact we could make with our loved ones was all the more important. Getting in touch with those we left behind was a tricky business. Unfortunately the ship's satellite telephone did not work for most of the trip. We were luckily in touch with Meri Laesk from Bluff, New Zealand, who operates a public service called Bluff Fisherman's Radio. Meri uses her short-wave (ham) radio to communicate with about 100 fishing boats every night. She was our connection to family and friends and would pass messages to and from, but more importantly, she was the safety link for each boat. We would give her our position in latitude and longitude, our weather and a report that all was well. If we failed to call in for several nights due to a serious problem, she could check her log and report our last known position to rescue authorities.

Meri's engaging personality enlivens you as she projects thousands of miles over the radio waves. It became a ritual that gave me great pleasure every night after dinner, when I would go to the bridge with Captain Iain Kerr to radio Meri. We became one of her Antarctic "iceboats." She checked with us just after 9 p.m. every night on a high frequency radio band, which had a good chance of bouncing from New Zealand to Antarctica, but which distorted voices so we sounded like we were calling from the bottom of a well.

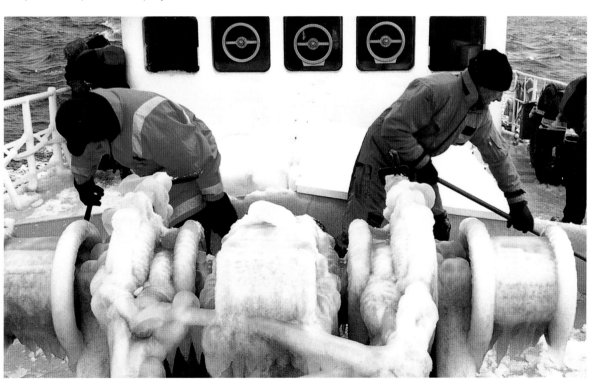

Top: The desalinator we used for making fresh water out of sea water failed halfway through our voyage. Here, Porter and I were collecting iceberg pieces to melt for drinking water, which actually tasted much better.
Photography: W. Skiles

Bottom: Ice on deck can cause the ship to become top-heavy and unstable, so the crew was vigilant in clearing the ice off with hammers and clubs.
Photography: J. Heinerth

27

After transmitting our position and weather to her, Meri would chirp back to us a cheerful, "Roger, Roger, *Braveheart*, that's good to hear, good as gold. Now you have a good night and I will see you tomorrow." Meri's simple message, "Greg sends his best and the boat is fine," gave enormous comfort to those at home and peace of mind to us as *Braveheart* lurched and pushed farther south towards B-15.

More than 1,000 miles and 10 days after leaving Campbell Island, we reached the icy waters of the Ross Sea.

Top: Occasionally, foggy conditions made icebergs impossible to see until we were right on top of them, so we relied heavily on the ship's radar.
Photography: P. Turnbull

Left: Pack ice has always been an obstacle to Antarctic exploration. Here, Porter and I discuss our options on *Braveheart's* bow as we slowly push through six-foot thick sheets of ice.
Photography: W. Skiles

Facing page left: We posted lookouts aloft to find leads in the ice. It was a freezing, but necessary, task.
Photography: G. Stone

Facing page right: Without Laurie's helicopter to scout leads through the ice, we would not have made it as far as we did.
Photography: P. Turnbull (taken from helicopter)

Braveheart approaches "Ice Palace," where we tested all the ice-diving gear, and made our first dives into Antarctic waters.
Photography: G. Stone

Facing page: *Braveheart* negotiates a tight spot among the giant icebergs of the Ross Sea. The ice pictured here is part of one enormous iceberg (about one-half mile long) connected by a submerged piece of ice.
Photography: W. Skiles

31

A small berg, named "Patience Camp" by the crew after one of Shackleton's encampments 85 years before, was a convenient mooring for testing ice-diving equipment.
Photography: C. Olavarría

Top: An aerial view of the iceberg we christened "Ice Palace." *Braveheart* is anchored to the berg.
Photography: W. Skiles

Bottom: To celebrate arriving at the Ross Sea, we set up camp on "Ice Palace" and barbecued steaks for dinner.
Photography: P. Turnbull

As Cold As It Gets

CHAPTER 4

The two high points of any Antarctic expedition are the first day you see ice
and the last day you see ice.

*"I will not say it was impossible anywhere to get in among
this ice, but I will assert that bare attempting of it would be
a very dangerous enterprise and what I believe no man in
my situation would have thought of. I whose ambition leads
me not only farther than any other man has been before
me, but as far as I think it possible for man to go, was not
sorry at meeting with this interruption, as it in some
measure relieved us from the dangers and hardships,
inseparable with the navigation of the southern
polar regions."*

-Capt. James Cook, *Journals,*
Voyage of the Resolution *and* Adventure (1774)

It has been said that the two high points of any Antarctic expedition are the first day you see ice and the last day you see ice. Ice is the defining feature, the one thing that you think about each day, that you see everywhere, that you hear scraping along the sides of the hull while you sleep, that you feel bumping off the bow while moving through the pack, that you melt for cocktails at night and that you worry about hitting in storms and darkness. It dominates every day in the Antarctic because ice IS Antarctica. Ice is what we came to study, ice in all its aspects, but primarily ice embodied in the phenomenon that was B-15, the largest iceberg in recorded history.

Antarctica is both the highest and the lowest continent because of its tremendous quantity of ice. More than 99 percent of Antarctica is covered by an enormous ice cap that averages 5,905 feet in thickness; Dome Argus marks the thickest ice, at more than 13,123 feet. Although the tallest mountain, Vinson Massif, is 16,067 feet, the continent's mean surface elevation is 6,562 feet, much higher than Asia, which, at 3,150 feet, is the next highest continent. It is the continental ice dome that makes Antarctica so high.[1]

The same ice that gives Antarctica this tremendous elevation also lowers the continent with its unimaginable weight. Antarctica's ice weighs about 40 quadrillion tons (400 Antarctic ice sheets equal the weight of the moon[2]), so heavy that it flattens the South Pole and makes the earth slightly pear-shaped.[3]

Facing page: "Ice Palace"
was an old iceberg
sculpted by years of
intense winds and
waves.
Photography: G. Stone

Most of Antarctica's land is actually depressed below sea level. If the ice were removed, the land would spring up (over thousands of years) above sea level. Antarctica's 7 million cubic miles of ice hold approximately 80 percent of all the surface fresh water on our planet, and about 90 percent of the freshwater ice. If the Antarctic ice dome was to melt, the world's oceans would rise about 200 feet, putting the roadbed of San Francisco's Golden Gate Bridge underwater.

The Antarctic ice dome is the result of hundreds of thousands of years of ice crystals lightly raining down from high altitudes over the pole.[4] Because the continent is so cold, the accumulation each year is more than the modest melting that occurs in the brief summer. As new ice is added, the dome grows and flows outwards, much like pancake batter on a griddle. When it reaches the edge of the land, the ice pancake flows out onto the ocean forming ice shelves, massive floating sheets of ice. Some ice shelves are several thousand feet thick. The Ross Ice Shelf is the largest of all, and it is from this that B-15 calved after its slow creep across the continent. Although B-15 didn't make the newspapers until it broke off from the Ross Ice Shelf, the ice that makes up the berg started forming about 75,000 years ago.

In addition to all the freshwater ice in Antarctica, there is also sea ice which begins to form when water temperature falls below 28.98 degrees Fahrenheit (-1.8 degrees Celsius). This frozen water, which is usually three to six feet thick, reaches out from the edges of the freshwater ice shelves and the continent across the Southern Ocean in an annual cycle of freezing and thawing. Each year more than 6 million square miles of surface ocean waters freeze and melt around Antarctica. That is an area larger than the Antarctic continent itself (2.7 million square miles) and about the size of Russia. The sea ice is a major force in the area. It controls the primary production of photosynthetic algae, the basis of the Antarctic food web; it provides haul-out sites for breeding seals; it limits the times some penguins can access the sea for food to only the summer months; and it forms edges of tremendous biological activity.

After a few days in the Ross Sea, we realized that scattered all around us were pieces of B-15. During 2000 and 2001 the Ross Sea was littered with pieces of this giant iceberg.
Photography: P. Turnbull

ABOUT ICEBERGS

We came to study the effects of ice on the ocean, primarily the ecology of large tabular bergs. Tabular bergs, of which B-15 was a giant among giants, take their name from their resemblance to large, flat tables. There are small tabular bergs in the Arctic, but the Arctic lacks ice shelves large enough to spawn giants in the same class as B-15.

According to the New Zealand Nautical Almanac, icebergs have six officially recognized sizes. "Growlers" are the smallest, about the size of a car. Growlers are dangerous because they can roll in the waves, just underwater, and are difficult to see. The ice is extremely hard and can severely damage a ship. From growlers they increase in size to "bergy-bits," which are the size of a small house. Bergy-bits are easy to see in daylight, but are a real problem if you hit one at night or in heavy weather. Next come the self-explanatory groupings: "small," "medium," "large" and "very large" bergs.

B-15 calved from the Ross Ice Shelf on March 17, 2000. About 4,500 square miles and as flat as a frozen pond, B-15 was a very large berg, the size of Jamaica or Connecticut. It contained more than a thousand cubic miles of pure freshwater ice, enough fresh water to supply the United States for five years. At the time of its birth it was the largest moving object on the planet. By the time we arrived in the Ross Sea ten months later, B-15 had broken up into two big chunks (B-15A and B-15B) and hundreds of smaller pieces.

IN THE ROSS SEA AT LAST

After pounding our way south through the entire Southern Ocean, which stretches from New Zealand for two thousand miles to Antarctica, we entered the Ross Sea on January 28 and found a whole new world. The sun shone 24 hours a day, and it was difficult to sleep without the diurnal light patterns that trigger sleep hormones. The air was so cold that spray off the ocean would freeze instantly on the rails and on the deck. The constant brightness of white sheets of ice and brilliant sunlight was broken up by occasional patches of dark blue water. Sound was different, muffled and quiet.

Top: Late night filming: Rather than rising and setting, the sun circled the sky, dipping lowest at midnight.
Photography: G. Stone

Bottom: The mountainous peaks of Cape Adare on the edge of the Ross Sea glow pink in the midnight sun.
Photography: G. Stone

The thrumming of the engines now seemed distant, almost like another ship's. I stood on deck hearing the crunching of the ice as it bounced off our hull, and saw the surprised look of crabeater seals *(Lobodon caracinophaga)* as they watched us slide through the endless fields of ice. Icebergs were everywhere with killer whales *(Orcinus orca)*, minke whales *(Balaenoptera acutorostrata)* and the occasional sperm whale *(Physeter macrocephalus)* swimming among them. After traveling so far and for so long, I suddenly realized we were here: We were surrounded by pieces of B-15.

Top: Laurie became skilled at landing the helicopter on small icebergs and ice floes.
Photography: P. Turnbull

Bottom: Crabeater seals are the most numerous seal species in the world. They are curiously named, as they feed primarily on Antarctic krill, not crabs.
Photography: P. Turnbull

NOTES FROM THE FIELD

Carlos Olavarría
University of Auckland, New Zealand. Molecular biologist and marine mammal specialist.
Photography: G. Stone

Right: We saw killer whales that may be a new species, smaller than typical and with unique markings.
Photography: P. Turnbull

I had been hoping to study the whales of the Ross Sea for five years when I was invited to be part of the *Braveheart* expedition. While studying humpback whales *(Megaptera novaeangliae)* in the Antarctic Peninsula several years earlier, I was surprised to learn that there were two kinds of killer whales *(Orcinus orca)*. One was the familiar type, but the other was yellowish and had a different pattern on its sides. Although I had seen just a few killer whales off of the Chilean coast, the first time I saw the yellowish kind, I tried to find out everything I could about this whale. With the help of friends, I read a Spanish translation of a Russian paper published many years ago in which the two forms were described. The yellowish one was, at the time, believed to be a new species.

Since the Russian paper was written, we have undergone a revolution of new molecular techniques that enable scientists to study the DNA of living organisms. I suspected that this could solve the mystery of whether these two types of killer whales were actually separate species.

Working in Antarctic conditions is not easy, to put it lightly. Just collecting a single sample is a challenge and I knew that I would have to leap at any opportunity. That moment finally came when a pod of killer whales swam alongside of us. I knew that I had to get the skin sample quickly, and that there might not be time for many attempts. The situation wasn't ideal because the group contained several calves, and I did not want to sample them or disturb their mothers. I decided to sample an adult male, hoping that the calves would stay with their mothers. I waited, looking through the laser view finder of my sampling gun (which doesn't do any damage to the whale, just takes a small sample), trying to place the red dot of the laser scope on the flank of the whale, following its movements under the water until it broke the surface. Finally, I got a good shot and got the sample. It was the first sample of an Antarctic killer whale from that part of the Ross Sea. Although it was not of the yellowish variety, I knew it would shed new light on the question.

After the trip was over and I was back at the University of Auckland, New Zealand, I found something exciting. I was able to sequence the mitochondrial DNA from the sample and compare my data against a global data set from that of my colleague, Rus Hoelzel, at the University of Durham in England. I found that our whale was one of the most distinctive killer whales in the data set, even though it had the common black and white coloring. We weren't able to answer the question of whether there are two different species of killer whales living in Antarctic waters, but as in all fields of science, the knowledge is built step by step. I hope this unique effort in one of the most remote regions of Antarctica will help the future work of others. While I also saw and studied humpback whales, minke whales and sperm whales on this voyage, the killer whale research was the most exciting.

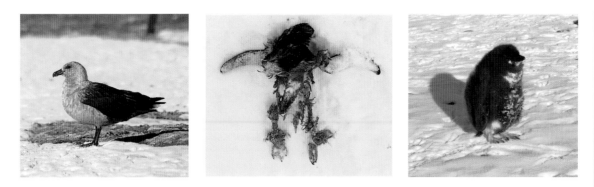

The large pieces were of particular interest. The biggest chunk, B-15A, had quickly come aground and stopped between the Ross and Franklin Islands, only about 150 miles west of the place on the Ross Ice Shelf where B-15 was born. As I write (summer 2002), B-15A is still there and may remain so for many years. It had a major ecological impact by keeping more than 170,000 breeding pairs of emperor (*Aptenodytes forsteri*) and Adélie penguins (*Pygoscelis adeliae*) from reaching their rookeries at Cape Crozier and Cape Royds. The berg blocked their way and added a two-day walk across the ice for the penguins from the water to the rookery. This increased the time they needed to forage for food in the sea and sadly resulted in only two percent of the season's chicks surviving.

The next largest piece, B-15B, was still adrift some 600 miles north in the Ross Sea when we arrived. Through its grand, slow-motion self-destruction, B-15 also created hundreds of smaller tabular bergs throughout the Ross Sea that provided ideal study sites for our science and photographic dives. Although they were not the "big one,"

they were still enormous and stretched out beside our ship as far as we could see.

Cloud-white with streaks of blue, sculpted along vertical faces like a diamond or a piece of marble, these massive bergs towered above us and extended below us like submarine cathedrals that disappeared into the depths. We could see but a small portion because only one-sixth to one-ninth of an iceberg floats above water level. Never before had I seen such monstrous objects floating in the sea.

UP CLOSE AND PERSONAL

How best to study them, now that we were really here? It was a tremendous science opportunity! I couldn't help but think of Pogo's [5] saying: "Some opportunities are so large they are insurmountable."

I knew we needed to approach as close to these huge bergs as possible and get into the water around them as much as possible. I wanted to learn how these bergs biologically and physically affected the surrounding ocean. What is the ecology of an iceberg? With more and more large bergs forming in this part of the world, it is important for us to understand their impacts on the ocean and their role on earth.

Top left: Penguin chicks are very vulnerable to attack from predatory birds such as this skua.
Photography: G. Stone

Top center: Mauled penguin chick carcasses were a common sight.
Photography: G. Stone

Top right: While we encountered Adélie penguins swimming and feeding throughout the Ross Sea, it was at Cape Hallett that we got closer looks at penguins and their chicks.
Photography: P. Turnbull

Bottom: Cape Hallett is home to more than 50,000 breeding pairs of Adélie penguins and an abandoned research station once jointly operated by the U.S. and New Zealand from 1957 to 1974. Small boat landings are allowed, but helicopter landings are not allowed because they disturb the penguins.
Photography: C. Olavarría

Cloud-white with streaks of blue, sculpted along vertical faces like diamonds or pieces of marble, these massive bergs towered above us and extended below us like submarine cathedrals that disappeared into the depths. We could see just a small part because only one-sixth to one-ninth of an iceberg floats above water level.

Photography: P. Turnbull

I first wanted to understand the water characteristics around the berg. To begin, we took some basic measurements: salinity, temperature and chlorophyll content. These are vital statistics of ocean water, much like a person's temperature and heart rate. Getting this information is one of the first things a marine scientist does because the results guide the next moves. I used a CTD machine for the job. About the size of a large thermos bottle, a CTD measures and records key physical and biological properties of water such as conductivity (C) which can be translated into a measure of salinity, temperature (T) and depth (D). The CTD also has a fluorometer to measure fluorescence, which indicates the chlorophyll content and the amount of plant life (phytoplankton) present, the basic building block of ocean food chains.

I was amazed to discover pools of fresh water around the bergs. Even in this extreme cold the icebergs were melting! Then I learned that the water actually got warmer as we went deeper. The freezing air temperatures were cooling the surface waters, while the deeper water was insulated from the freezing air. We also found increased phytoplankton near these giants. Maybe the icebergs were biological engines, churning the water around them and encouraging life to grow.

Top: We used an instrument called a CTD to measure salinity, temperature and chlorophyll levels in the water near pieces of B-15. We found that, even in freezing temperatures, the icebergs are melting and forming freshwater pools around themselves.
Photography: G. Stone

Bottom: I anxiously watch as a swell surges under my inflatable boat and nearly propels me into the jagged roof of an ice cave, interrupting data gathering.
Photography: W. Skiles

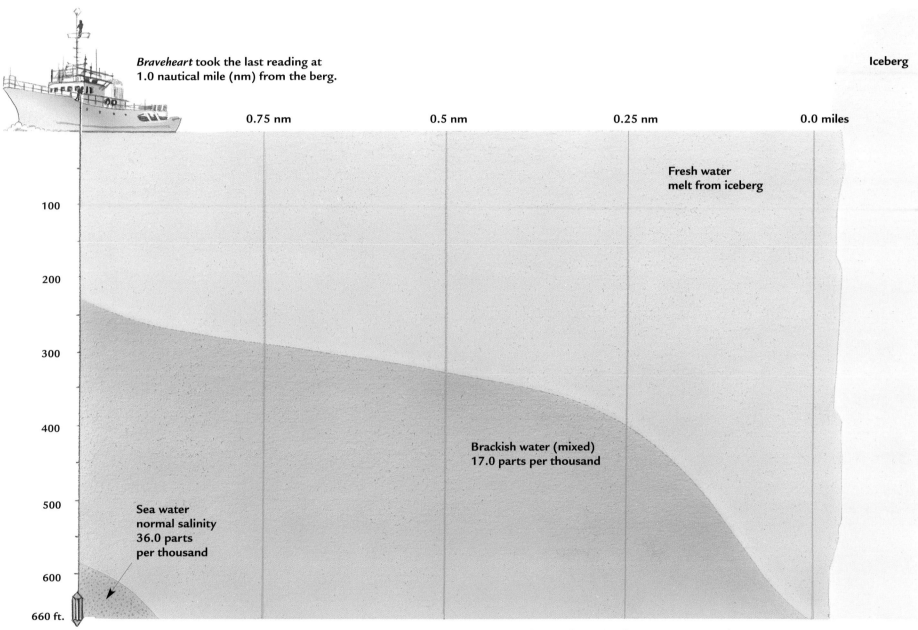

Braveheart took the last reading at 1.0 nautical mile (nm) from the berg.

Iceberg

0.75 nm   0.5 nm   0.25 nm   0.0 miles

Fresh water
melt from iceberg

100

200

300

400

**Brackish water (mixed)
17.0 parts per thousand**

500

**Sea water
normal salinity
36.0 parts
per thousand**

600

660 ft.

The salinity profile of a large berg (10 miles long by 5 miles wide) shows an unexpected pattern. The iceberg is melting even in this cold water, creating a wedge of fresh water around itself at the surface. Salinity is measured in parts per thousand: the number of parts of salt contained within 1,000 parts of seawater.
Illustration: B. Harmon

Now we needed to don our diving gear to learn what life was like in the water around and inside these bergs. Diving in Antarctica is supremely beautiful but incredibly, unremittingly freezing. To survive, we brought dive suits worthy of NASA. The near-frozen sea water sucks away body heat so quickly that an unclothed person would die in minutes.

Suiting up started with strapping electric heating pads over our kidneys with duct tape. The kidneys are considered the most effective area to heat, since blood circulation goes deep into the body at that point. Even with this heating device our toes and fingers got terribly cold. Then we put on heavy duty polypropylene underwear. After that we climbed into thick quilted thermal underwear made especially for diving (retailing at around $400). Finally, we got into the outer shell of our dry suits, made of reinforced rubber. We were justifiably paranoid about leaks in this aquatic armor and would meticulously examine all the wrist and neck seals to be sure they were sealed properly. Provided the suit did not tear, the wrist and neck seals were the only places water could get in. During one dive, Wes's suit leaked badly from a faulty wrist seal. By the time we got

Top: Dive preparation was a team effort. Matt, Nigel and Keith Moorehead help Porter and me put on our suits, test seals and prepare to enter a freezing alien world.
Photography: C. Olavarría

Bottom series: Getting in was the easy part. We needed help hauling ourselves and our 100 pounds of gear onto the ice after a numbing and exhausting dive.
Photography: C. Olavarría

him out of the water and back in the boat, he was hypothermic and could not stand up. We carried him to the shower room where we stripped off his gear and bathed him in hot water until he could move again.

The gear added so much bulk and buoyancy that we needed about 50 pounds of weight to make ourselves heavy enough to submerge. After an hour of dressing and preparing our equipment, we staggered from our dive locker across the ice-spattered deck of *Braveheart*, trying not to slip, to the skiff that would take us close to the berg. Matt Jolly was our skiff driver and the best dive assistant in the world. He would help us complete our armor by fitting the all-important hoods around our necks and the gloves to seal our wrists as we headed out toward a mountain of ice. Perched on the side of the boat, we tried to get up the guts to slide in. Taking one last gulp of air and slamming the regulators in our mouths, we plunged into the bone-chilling water that sought every tiny piece of exposed skin and tested each seal.

Top: The ice is freezing around me as I submerge into the sea ice of the central Ross Sea. My regulator is out of my mouth to prevent it from freezing in the frigid air. We always waited until we were underwater before taking our first breaths.
Photography: C. Olavarría

47

NOTES FROM THE FIELD

Porter Turnbull
Expedition medical officer, ornithologist and science diver.
Photography: C. Olavarría

Right: Greg and I were excited about doing our first dive at our "Ice Palace" research site.
Photography: W. Skiles

When Greg asked me to literally follow him to the end of the earth, I did not hesitate to accept. The opportunity to dive with him under the ice in Antarctica epitomizes adventures we have shared over the years. The risks and dangers involved in such an undertaking were understood and assessed along with the scientific objectives of the expedition. But my most important role was realized in a brief encounter with Austen, Greg's wife. She had a few questions. Yes, the ship was very small. Yes, we would be voyaging the stormiest seas of the planet. Yes, diving in this frozen sea and every aspect of our journey was dangerous. Then, while Greg was out of the room, I promised her that I would watch out for her husband, one of my best friends, and see him home safely to her.

Upon reaching the Ross Sea, our diving challenges began. The gear was so cumbersome that we needed help to place our masks on our faces and regulators in our mouths. Greg and I would inspect each other before immersion: All gear in place, not a bit of flesh showing. These pre-dive preparations typically required an hour or more. The buddy system was the order of the day. Greg's signals to me were emphatic; in staccato gestures he would point to me, to his eyes and then to himself (you watch me!). I found myself remembering my promise to Austen and at times I would wait and make sure I was obviously watching him when he turned my way. When we were on air he was never out of my sight. The formation of ice crystals in our air supply would result in sudden free flowing of our regulators. These events required us to switch to a spare air bottle while the other diver shut down the valve of the primary tank. To prevent regulator freezing, we soon learned to dry our airlines with pure nitrogen and not to open our air valves until after we were totally immersed.

One thing that fascinated me was the adaptive physiology of my body. After repetitive dives I learned to control the flow of warming blood to my extremities. I found that a type of biofeedback response could reverse my body's initial vasoconstrictive reaction to cold immersion. This willful psychosomatic response allowed for vasodilation and the resulting warming of my hands and feet helped ease the pain in the freezing waters.

Despite all the technical considerations and dangers, diving in the middle of the Ross Sea beneath pack ice, as well as alongside cathedral-like icebergs, was one of the most spectacular underwater experiences of my life. The presence of leopard seals (*Hydrurga leptonyx*) and other apex predators, like killer whales, added to the neurosensory overload. Each dive was an exhilarating adventure with the promise of scientific and self discovery.

Sea water freezes at 28.98 degrees Fahrenheit (−1.8 degrees Celsius) because the salt content slows the freezing process. Most of the water we dived in was 28.99 degrees Fahrenheit (−1.73 degrees Celsius). Sometimes we swam through water that was a slurry; it was freezing while we were in it. The difference between the freezing point of sea water and fresh water meant that any fluid with lower salt content than the sea water could actually freeze while we were diving. These other fluids included our blood and the fluids in our skin cells. There was the very real potential for frostbite during an hour of submersion. We also had freezing regulators. If there was any small amount of fresh water in our breathing regulators, the water would freeze and air would burst at high pressure into our mouths. It was nearly impossible to breathe this high pressure air (like taking a drink from a fire hose), and it didn't help that the air was at sub-zero temperatures because it had been in the scuba tanks on our backs.

If the freezing air blasted into our mouths for too long, it could freeze and crack our teeth. This had happened to other divers in the Antarctic recently and, as we were all fairly fond of our teeth, we were careful to avoid this. The water was so cold, but the air was even colder; if we took a breath before entering the water, the regulator would sometimes freeze from the water vapor in our breath even before the dive began. Despite all the layers and precautions, I would float at the surface and moan for several minutes before numbness blessedly crept over the world's worst "ice cream" headache.

Then the magic began. All around us were white and blue walls of ice stretching down into the deep dark. The water around us shimmered from the cold waves, like images on a hot desert day. The water was as clear as some tropical dive sites — like gin. Suspended, I felt at home and exhilarated. Even with all the problems and the cold, Antarctic diving is still my favorite.

Top: Jill and Paul submerge with rebreathers to photograph and film the animals living around this iceberg. Rebreathers help keep divers warmer by recycling each breath, removing carbon dioxide and enriching the loop with oxygen. Porter and I used open circuit scuba regulators and tanks, but got chilled to the bone from continuously breathing the freezing air and we had trouble with freezing regulators. *Photography: W. Skiles*

Bottom: We used a Remotely Operated Vehicle (ROV), an aquatic robot, to explore and film along the edges of large bergs at depths too deep for divers. We strapped our CTD on the ROV to collect data on temperature, salinity and chlorophyll content near and inside these giants. *Photography: P. Turnbull*

49

THE FIRST DIVE GOES WRONG

Our first dive started well and then went terribly wrong. We were 300 miles from shore in the central Ross Sea, but the closest land was 6,000 feet below us — the deep-sea bottom. We had found our first iceberg, a remnant of B-15. Porter and I had trained and rehearsed this occasion many times on the way to Antarctica. We knew this was the most dangerous diving in the world and we were ready for every scenario except what actually happened.

We vented the air from our buoyancy compensators and began to descend with our 50 pounds of lead weights pulling us into the deep. Off to our right we saw the wall of ice stretching down into darkness and out of sight. Above we could see the brightly lit day and the silhouette of our red skiff with Matt's wool-capped head looking over the side at us as we descended. All we could hear was the whooshing of air through our regulators.

Looking at Porter, I pointed to the wall and motioned that we should swim towards it. Then the trouble began. As we neared the wall, we began sinking too rapidly. We kicked up but could not stop the descent. The wall of ice was now rushing past us, as the sunny day and outline of our boat above got smaller and smaller. I looked over and saw Porter's eyes wide and intense. We were at 80 feet and still sinking. We channeled air into our dry-suits and buoyancy jackets, but nothing slowed us.

Finally, at 100 feet we began to rise and regain control of our dive. We moved away from the berg and realized we had been caught in a density-driven downdraft at the berg's edge.

DISCOVERIES

Our first discovery was that large icebergs in the open ocean cool the water around them, creating powerful down-currents at their edges, too strong for divers to fight. Later, we used fluorescent dye to measure the current of that water. By releasing the dye and videotaping its movements, we were able to measure the current at more than one knot (about 1.15 miles per hour)![6]

After getting caught in several more downdrafts and dealing with a series of frozen regulators, we mastered the necessary techniques and began our work in earnest. We discovered that juvenile icefish (Nototheniidae) live in holes in the sides of the iceberg. Their ice holes provide protection from predators like seals and penguins and help the fish pass successfully through their vulnerable juvenile stage. We could not determine whether the icefish excavate the holes or use existing holes, but we did learn that some of these fish were living in fresh-water pools at iceberg edges, a new observation for this species.

Antarctic icefish have adapted in remarkable ways

pulse and drift through the water hunting for food with stinging tentacles. These beautiful lacey creatures feed on small Antarctic krill (*Euphausia spp*.) and are in turn fed upon by snow petrels (*Pagodroma nivea*) and other birds. One day in the middle of the Ross Sea, we swam through a dense patch of these siphonophores, finding them suspended in the water column every ten feet in every direction in a three-dimensional matrix. When petrels began diving in and feeding on the siphonophores around us we glimpsed Antarctic ecology in action: The birds nested on the iceberg while the iceberg created oceanographic conditions beneficial

because of the physiological gymnastics required to survive in their home. Some have evolved adaptations to the extreme environment such as low metabolic rates and high rates of protein synthesis. They can achieve neutral buoyancy through changes to their skeletons and body lipid content, enabling them to remain in the mid-water zone where tiny shrimplike krill, a primary food source, live. Other icefish have a type of "antifreeze" in their blood called trypsinogen, a glycoprotein produced by the pancreas, that keeps them from freezing. These fish are the only vertebrates that lack hemoglobin, so their gills and internal organs are colorless. Our observation that they can also live in fresh water, at least during juvenile stages, is another interesting adaptation.

We found clusters of red siphonophores (Agalmatidae) which are three-foot-long jelly-like, boneless, colonial creatures that

Top left: This Antarctic siphonophore was abundant and provided food for snow petrels in the waters near several icebergs. *Photography: G. Stone*

Top right: Krill (*Euphausia superba*), one of the most abundant animals on earth, can live to be seven years old. The green in its gut is recently eaten phytoplankton. *Photography: G. Stone*

Bottom: To measure the current flow around an iceberg, I released a yellow biodegradable dye that allowed us to track water movement on videotape and then calculate its speed. *Photography: W. Skiles*

51

to the siphonophores. The iceberg was the central character around which all this activity was based. Sea jellies inhabited the waters near icebergs at higher densities than in the waters farther from bergs. We found healthy populations of not only *Beroe*, a genus of ctenophore, but also numerous Cydippida — smaller ctenophores often preyed upon by *Beroe*. We found about one *Beroe* for every five to seven Cydippids, a normal predator-prey relationship, associated with the edges of these icebergs.

Two other fascinating features of Antarctica's ice were the dense swarms of shrimp-like krill and the mats of phytoplankton and tiny seaweeds covering the underside of the ice. This plant material freezes into the ice, providing food during the Antarctic winters. It was clear that these monstrous islands of ice, with the surrounding sea ice, were creating their own ecological systems in the waters near them.

Our study of bergs progressed well as *Braveheart* pushed her way through the ice of the Ross Sea. We fell into a routine of making one dive a day, and organized it to sample as many different ice conditions as we could find on that day. One dive a day under those conditions was all that we could tolerate. After each dive we tried to warm up by taking hot showers and wrapping ourselves up in our sleeping bags. After the initial cold diminished, we practically inhaled large quantities of food in the galley and then slept for an hour or so, snugly wrapped in our sleeping bags again. While we dived, Carlos continued his whale studies with helicopter and vessel surveys.

Midway into our trip, a major objective still eluded us. We wanted to dive around B-15B, the second-largest piece of B-15 and one that no person had yet visited. We hoped to bring *Braveheart* alongside the giant to take close-up video and photos and to observe whether there were strong peripheral currents like those we had found with B-15's smaller siblings. Sadly, we soon realized it would be impossible to reach B-15B and return home given our fuel and provisions restrictions. The pack ice surrounding B-15B was just too thick to navigate in the time remaining. We later learned it was a particularly bad year for pack ice in the Ross Sea. Disappointed that we wouldn't achieve that goal, we decided on an alternative – someone would fly to B-15B in the expedition's tiny helicopter.

Porter holds a net designed for catching marine plants and animals for study.
Photography: G. Stone

52

Top: A Cydippid ctenophore, one type of comb jelly we found living around the icebergs.
Photography: G. Stone

Second: These *Beroe* jellies prey on Cydippid comb jellies, shown above.
Photography: G. Stone

Third: *Glyptonotus antarcticus,* an Antarctic isopod found living beneath grounded icebergs and able to withstand frigid water.
Photography: G. Stone

Bottom: We found this isopod and sea urchin *(Sterechinus spp.)* under a large iceberg.
Photography: P. Turnbull

Taking It to the Air

Wes and Laurie agreed to try landing on the berg, with Laurie piloting the 40-mile flight over pack ice separating *Braveheart* from B-15B. The helicopter was not designed for such a long, sustained flight over water. Its main purpose was to hover near *Braveheart* to scout openings in the pack ice and to ferry people and gear short distances from the ship to icebergs.

Laurie and Wes made special modifications to the tiny Hughes helicopter, strapping extra fuel to her outside and storing emergency gear like tents, food and sleeping bags into every available space. It was a risky flight and we all worried that they might not make it to B-15B and back safely. Wes and Laurie filed a flight plan with me, which allowed them four hours for their quest, including three hours to complete the trip and one hour contingency. The iceberg was just beyond radio contact, so if we did not hear from them in three hours, we would engage an emergency protocol for their rescue.

Landing a helicopter on an iceberg is not easy. Laurie would have to be ready to take off immediately in case the chopper landed on a deep crevasse hidden by fresh snow. If the helicopter tumbled into a crevasse, we would never find them.

Fortunately, their flight went flawlessly and they were the first people to land on and bring back firsthand observations of B-15B. As they told us afterwards, Laurie carefully landed his tiny craft on the iceberg and Wes ventured out and walked across to the iceberg's edge. Wes returned with astounding video and still pictures, making essential observations along the base of B-15B where there were strong currents and abundant marine life.

The Underbelly of an Iceberg

Marine scientists have little biological information about what is beneath icebergs and what may live deep inside them. To find out, Paul and Jill planned to use their rebreather diving gear for research surveys to these more remote parts of an iceberg. Both are expert cave divers and were eager to do something I was not – to dive into and under one of these giants. So, we looked for a good one. We left the ice-choked waters of the Ross Sea and headed for a coastal environment for this last exploration. We dived off Cape Hallett for about a week to characterize the region and find an appropriate iceberg. This was our first opportunity to explore the seafloor life of Antarctica. All of our previous diving had been in the open ocean around drifting icebergs.

Trapped in pack ice, we were locked tight 40 miles from our destination. Since we couldn't reach it by boat, Laurie and Wes flew to B-15B in the helicopter. It was a risky flight, and we all worried about their safe return.
Photography: W. Skiles

54

barrage of icebergs drifting into Cape Hallett. Most of the animals were filter feeders, meaning they each had ways of filtering tiny plants, animals and organic material from the water for food. They grew well under the berg because it channeled a fast flow of water giving the animals a continual smorgasbord of nutrition on each tide. It was a Garden of Eden in one of the most inhospitable places on earth. From the amount of growth on the bottom, it looked like the berg had been there for three to five years. The berg had been in one place long enough to create its own habitat and a community of benthic creatures (bottom-dwellers) to inhabit it.

At Cape Hallett, we found the sea bottom to be covered in massive kelp, but with few animals. There were jellies, siphonophores and icefish in the water column and close to the icebergs, but the sea floor was barren of life. It was furrowed like a cornfield plowed by a dizzy farmer, from icebergs crashing in and coming aground on the bottom. Sand and gravel were piled in uneven rows in all directions. I realized that Cape Hallett was like a billiard table pocket, catching drifting icebergs from the Ross Sea, including the one we were about to explore.

We chose a grounded iceberg, assuming it would be stable and relatively safe for the high-risk venture. Porter and I conducted a series of dives around the edges and then Wes, Paul and Jill dived under and inside the iceberg, with Wes filming using his high definition camera.

To my astonishment, the trio discovered that this iceberg harbored and encouraged a dense growth of sponges, sea cucumbers, feather duster worms and other marine life that normally lives on the sea bottom. There was a rich carpet of life growing on the sea floor beneath the iceberg. The grounded berg apparently provided necessary shelter from the constant

How long would the berg stay there? How long would such abundant life find refuge? We soon found out.

Top: The elusive B-15B, which we only reached by helicopter.
Photography: W. Skiles

Bottom: We lit flares so the helicopter wouldn't lose *Braveheart* in the vast whiteness.
Photography: G. Stone

55

NOTES FROM THE FIELD

Jill Heinerth
Explorer/diver, with
husband Paul.
Photography: W. Skiles

Top: Paul enters the
water to explore the
underbelly of an iceberg.
Photography: J. Heinerth

Middle: Diving in water
as cold as water can get
and still be called water,
Jill led the way into a
broad crevasse.
Photography: W. Skiles

Bottom: The bottom was
rich with sea cucumbers
(*Abyssocucumis turquet*).
Photography: W. Skiles

We cautiously entered a collapsed area to find a gaping fissure that extended out of sight. Sheer white walls dropped steeply into a narrow crack. We swam into the fracture a good distance and drifted down to the sea floor. As we hit 130 feet, we discovered that the berg was undercut and we could continue our swim below the mass. We found a dazzling world of colorful tunicates, sea stars and curious creatures. Brilliant reds of the sea life cast a glow on the underside of the ice just a few feet over our heads. The allure of the teeming life entranced us to swim beneath the great berg and explore the expansive cave environment. This environment was completely different from the scoured sea floor we had seen so far. We slipped silently through the underbelly of the berg with our closed-circuit rebreathers, hearing only the occasional fire of the solenoid valve, a comforting click-hiss that means oxygen is being added to the loop. We found tiny columns and ridges that fixed the berg in position. The forces of great currents carved conduits and passageways through the ice and brought nourishment to the plentiful life. Large scalloped hollows textured the walls like dimples on a giant golf ball.

The current accelerated at a horrifying rate during the dive. When we turned the dive, it was because of the torrent of water that bore down on us. Getting out of the cave became a frightening fight for our lives. There was nothing that enabled a handhold. We needed to get off the floor but the walls were slick ice. Small icefish holes were the only offering that gave us an opportunity for rest. They were just large enough to insert a single digit. I moved up, finger by finger, evicting the tiny fish out of their dens, sending them into the siphoning cave. Paul helped Wes drag the massive camera and we finally got in the lee of an ice formation at 90 feet.

Our rebreathers had given us the luxury to breathe hard without failure. Our extra gas supplies and the ability to dive with optimized oxygen in the loop had given us the comfort of time to solve our problems. We were cold and tired after our decompression, but at least we were alive. A diver in open circuit gear would not have survived the three-hour ordeal. He would have run out of gas or suffered from irreversible free flows. If the equipment didn't kill him, then the tiring decompression would have ended with hypothermia.

THE ICEBERG VANISHES

That same night, I was awakened by shouts of "It's falling apart! Get up here, quick!" It was 1 a.m. and as I ran from my cabin to the deck of *Braveheart*, the Antarctic sun glowed red on the horizon. All 18 expedition members were gathered on deck, transfixed, as the enormous iceberg we had been studying heaved upwards, one end pausing high in the air like the bow of a foundering ship, then crashed down, creating waves that swept through all of Hallett Bay and rocked our boat.

What happened next none of us could have imagined—the iceberg that we had chosen for its stability, the one that was grounded and home to abundant life, rose one last time and seemed to explode into millions of pieces like shards of crystal, covering two square miles of ocean. Later, we circled the debris field of shattered ice. We had witnessed the violent end of an iceberg, the cataclysmic moment when its internal structure melts to the point it can no longer support its own weight.

We had moved our boat only two hours before the berg flipped and exploded. We had no reason to move the boat other than wanting a change of scenery that night. A lucky decision.

Why that iceberg disintegrated in such a spectacular fashion, none of us knew at the time. I later consulted with glaciologists and learned that throughout its life, an iceberg is in constant thermal change. Each day in the relative warmth of the Antarctic summer, the iceberg melts a little. Then each night it refreezes. This cycle created fissures and latticework throughout the ice. This is exactly what Jill, Paul and Wes saw during their dive into the giant. But then a cataclysmic moment comes when there are so many internal cracks and fissures that the berg can no longer support its own weight. At the coldest moment of the day (1 a.m.), when the nightly refreeze began, the giant irreversibly began to crumble. It was one of the most dramatic moments of my life and was a scene perhaps never before witnessed on such a scale.

The precariousness of our position didn't sink in until the next morning when we awoke to another sign that our time in the region was near its end. The ocean was freezing all around us. It was February 23 and the Antarctic summer was rapidly waning. The sun began dipping below the horizon each night for a few minutes and the winds had a colder bite. The sea was forming pancake ice, the first stage of the annual freeze. Once the pancake ice begins, it will advance nearly three miles per day until it reaches more than 700 miles out from shore. We were all weary; many of us suffered marginal frostbite and had lost feeling in our toes and fingertips from neuralgia due to the constant cold during dives. The exploding iceberg was a shock, and the reality of where we were on our planet was emphasized by the freezing ocean. It was time to go home.

Left: This enormous iceberg, which we had been studying only hours before, suddenly heaved upwards, one end pausing high in the air like the bow of a foundering ship, then crashed down, creating waves that swept through all of Hallett Bay and rocked our boat. And then it exploded. *Photography: W. Skiles*

Right: After the iceberg exploded, what was left was shattered ice covering the water's surface and an open wall where once there had been a crevasse. *Photography: W. Skiles*

With pancake ice forming around us, *Braveheart* turned her bow north and started back through the ice and across the Southern Ocean. Pancake ice is the first stage in the annual freezing of Antarctica's ocean waters. *Braveheart* needed to leave the Ross Sea before the sea froze solid.
Photography: G. Stone

HEADING NORTH AGAIN

Braveheart turned her bow north that morning and started back through the ice and across the Southern Ocean. An old saying among mariners helps explain our voyage home: "It is far easier for a ship to take it on the ass than on the shoulder; easier for a ship to run than to punch." While the ocean we re-crossed had not changed, our orientation to the waves was better going north. The gigantic, persistent waves of the Southern Ocean came more from our stern than on our bow during that return voyage. Still, we were again wedged in our bunks or holding tight to *Braveheart's* steel railings as we pounded, lurched and rocked our way home.

The return voyage was easier for another reason. We did not have the unknowns of Antarctica's ice fields and B-15 looming ahead of us. Instead, we looked forward to the green hills of New Zealand and to seeing our families again. Because there are no terrestrial plants in Antarctica, we had not seen a green leaf for two months. It was delightful to feel the air get warmer with each passing mile.

The food did hold out, though we ran out of coffee and tea some five days from port. We were all on deck, back in our summer t-shirts and bathed in sunlight the day we docked in Akaroa, New Zealand. The smell of farms and trees filled our senses and a group of Hector's dolphins swam around our bow, as if to welcome us home.

Top: Pack ice, as pictured here at Cape Hallett, is frozen sea water, while an iceberg is frozen fresh water that has broken off from a glacier.
Photography: P. Turnbull

Bottom: As we started making our way back home, *Braveheart* got trapped in thick pack ice for another day. We used the crane and platform to lower people onto the ice to explore the surrounding area.
Photography: G. Stone

CHAPTER 5

A Global Warning

The large tabular icebergs of Antarctica are biological engines.

It will take an oceanographer to find life in space. NASA has now concluded that the most likely place to find off-Earth life in our solar system is on the sea floor of the only other known ocean on the small moon Europa circling Jupiter. It appears that Europa has a very deep (more than 18 miles) salty ocean, frozen only at the surface. NASA is now working with the oceanographic community to plan a mission to explore the ocean of Europa for the presence of life. As we gaze into the heavens in search of life, we first look for an ocean because of its life-giving properties.

But are we taking care of our own ocean here on Earth?

Landing on the warm New Zealand shore, we were elated to have completed our journey safely and even more excited to take a good look at our findings and analyze them — to put the puzzle pieces together and figure out what it all means.

On our expedition we learned a great deal about the ecology of large icebergs. We determined that the large tabular icebergs of Antarctica are biological engines, sheltering and perhaps enhancing the rich biology found along the edges of ice shelves. Throughout the oceans, the biological action is usually found along edges of currents, islands, continents and large objects. These edges help concentrate and mix nutrients required for fertilization of the oceans and the growth of tiny single-celled marine plants called diatoms, the basis of marine food chains.

Icebergs, too, act as edges. As they drift and melt, large icebergs create powerful currents that leave physical and biological wakes in the ocean around them. These wakes may encourage the concentration of krill, fish, sea jellies, whales and seals. The bergs also create habitat for the icefish living inside and underneath, for penguins and seals climbing up on ice ledges, and for seabirds nesting on them.

When Wes, Paul and Jill dived in and under the berg that later disintegrated before our eyes, they learned what lives beneath a giant berg. The fissured latticework and caves and the abundance of worms, sea cucumbers, sponges and other invertebrate life, fed by strong currents at the base of the berg, were important discoveries.

Our findings from this expedition also furthered our understanding of the effect of global climate change on the oceans and its sea life. Gigantic bergs and melting ice are symptoms of global warming that will have short- and long-term effects on the ocean and our planet.

Facing page: A southerly gale blows across a large piece of B-15 in the Ross Sea. *Braveheart* took shelter behind this 200-foot-high berg until the gale subsided.
Photography: C. Olavarría

Oceans — The Lungs of Our Planet

All the planet's life is dependent on the global ocean. Without the oceans' interflowing system of salt water, nothing could live. The "earth as a spaceship"[7] analogy is a useful way to evaluate our natural resources. As we hurtle endlessly through space with no chance of re-supply, what does Spaceship Earth have on board? We have a modest air supply, some fresh water, some land, and then we have the immense body of salty water that regulates all climate, produces most of the oxygen to replenish our air supply, and contains the majority of the other living creatures and plants. Our oceans — the very lungs of our planet — are the primary life support system for Spaceship Earth. They are also where most of the biological action happens, since they cover 70 percent of the surface area.

If you quantify the area on earth where organisms can live, from the bottom of the sea to the top of Mount Everest, the global ocean contains more than 90 percent of this volume, the livable space on earth. While we must take care of all parts of the earth, the oceans are the most critical, the most biologically rich and the most dominant feature. Where would we be without our oceans? We have a solar system full of dusty, hot and cold planets that are examples of what Earth could be like without an ocean. If extraterrestrial aliens ever do visit Earth in search of life, chances are that, like our planned exploration of Jupiter's moon Europa, they will look in the oceans first.

Thomas Malone, chief scientist for Sigma Xi and distinguished scholar emeritus at North Carolina State University, said of the twentieth century:

> "A thousand years from now this will be looked upon as the first century in the history of the earth when one species, ours, gained parity with nature in determining the future habitability of the planet for ourselves and the other living creatures."

Although we may not want to take responsibility for it, he appears to be right. We now have the ability to change the climate and degrade the life support systems upon which we and all other life depend. The harmful effects of pollution, especially synthetic chemicals, have been known for decades, but we are just beginning to understand climate change. There is little doubt that human activity, particularly the burning of fossil fuels (which increases carbon dioxide and other gases in the atmosphere), heats the planet through the "greenhouse effect" and is changing the earth's climate. We must learn to what degree we are changing the planet and what the consequences are for ourselves and other organisms.

To understand change in any system, we need to identify the indicator regions, those areas first to react to the new conditions and to show signs of change. First to show evidence of a warming earth will be the two regions of temperature extremes, the warmest and coldest parts. It will be harder to find change in the temperate zones where animals and ecosystems exist within a wider range of seasonal climatic and ecological conditions. The tropical and the polar ecosystems are where sensitivity is greatest and where we already find animals struggling and ecosystems destabilizing.

Top: Global warming is evident in the shrinking ice cover in both the Antarctic and the Arctic. Animals that rely on ice cover are already struggling. *Photography: W. Skiles*

Bottom: Adélie penguins have already suffered population declines as a result of global warming. *Photography: J. Heinerth*

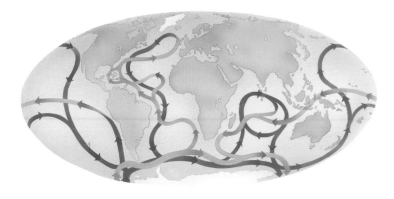

In the warmest areas, the tropical seas, coral bleaching has become a serious problem. Corals and the symbiotic algae that live in the coral tissue (zooxanthellae) are temperature-sensitive, and even a small increase in the warmth of their environment is having dramatic negative effects.

There is plenty of evidence of our warming planet at the poles. Arctic sea ice, the frozen sea water covering the area around the North Pole, has been thinning and shrinking for the past 50 years. Since 1978, an additional area the size of Texas has become ice-free. Average ice thickness has decreased 42 percent (from 10.2 to 5.9 feet) since the 1950s. Lack of sufficient snow cover is jeopardizing successful denning and reproduction in polar bears. On August 19, 2000, *The New*

York Times ran a front-page story reporting that an icebreaker full of adventure-tourists found a one-mile-wide ice-free patch of ocean at the North Pole. Scientists now predict that by 2050 the Arctic will be largely ice-free in the summer, making the Northwest Passage, the sea lane across the pole between the Atlantic and Pacific Oceans, a routine and predictable route for commercial vessels. The melting of the Greenland Ice Cap, the largest freshwater ice sheet in the world after the Antarctic Ice Dome, is also partly attributed to global warming.

Ecological symptoms of global warming have also been documented in the Antarctic Peninsula where glaciers are retreating and the seasonal cover of sea ice is shrinking, similar to what is happening in the Arctic. Temperatures in the Peninsula have risen 3-5 degrees Fahrenheit (2-4 degrees Celsius) in the past 50 years. The retreating glaciers have opened up more land sites for nesting penguins, but the reduction in ice cover has diminished winter krill populations. Krill depend on ice cover for protection from predators and for food, as they eat the phytoplankton living in and under the ice. Less sea ice means less food during the winter for krill, which means that fewer krill survive those long dark months. With fewer krill, penguin and whale populations may decline because krill is a staple of their diets. For example, a decrease in some Adélie penguin (*Pygoscelis adeliae*) populations in the Antarctic Peninsula region is attributed to global warming's effect on the whole web of Antarctic animal populations.[8]

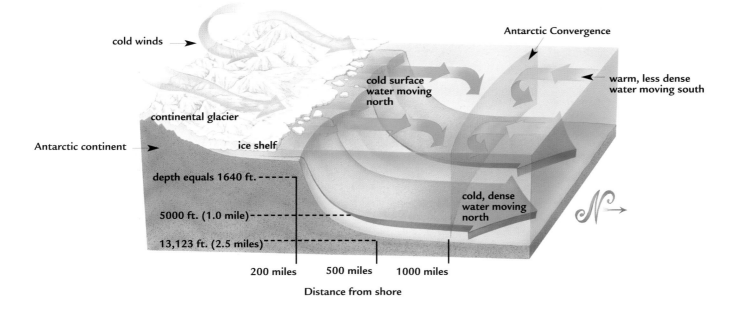

Top: Ocean currents are connected throughout the world and many are dependent on the chilling effect of Antarctica's ice and on wind to drive them. Dark arrows represent deep currents and light arrows represent shallower currents.
Illustration: B. Harmon

Bottom: Cold winds sweeping down Antarctic ice shelves help chill surrounding waters, which, in turn, help drive global ocean currents.
Illustration: B. Harmon

A Ross seal
(Ommatophoca rossii)
rests on pack ice.
Ross seals are the rarest
of the four seal species
that breed on Antarctic
pack ice. It is difficult
to study these animals
because of their widely
dispersed and isolated
distribution.
Photography: C. Olavarría

DANGER ZONE

Some might call melting ice and coral bleaching early warning signs. I would say we're already in the danger zone. Tinkering with our life support systems is the largest, most significant uncontrolled experiment human civilization has ever conducted.

Even more important than these short-term changes in ice cover and penguin populations are the potential long-term changes to ocean circulation. If oceans are the primary life support system on Spaceship Earth, then the Antarctic is one of the main engines powering this life-giving, life-supporting water system around the planet. The Antarctic Convergence (or Polar Front) is that area where the cold, northward-spreading surface current converges on the warmer waters of the lower latitudes. When they meet, the colder, denser Antarctic

water sinks and disappears under the warmer water mass. This convergence line is also the limit of the pack ice, and is generally considered the boundary between the Southern and Antarctic regions.

This sinking cold water is a major force in the global ocean current system and is called "Antarctic bottom water." The dense Antarctic bottom water forms an underwater stream that spills to the sea floor and flows north. Frigid waters of the underwater stream help drive the interlocking conveyor belts of ocean currents, the lifeblood of earth. The cold water eventually warms and rises back to the surface hundreds to thousands of years later in tropical latitudes.

An alarming possibility is that global warming could alter the deep sea currents. When sea water freezes, dissolved salt is left behind forming a cold, dense brine that tends to sink.

64

The deep current system is driven, in part, by this cold, salty, sinking water from Antarctica; as that water becomes diluted from the melting ice sheet, the salinity of the water decreases, making it less dense than the warmer waters to the north. If global warming continues and this ocean circulation driver weakens, we could change our climate more quickly and unpredictably, altering a planet-wide system and potentially making some areas much colder and others warmer.

There is a staggering amount of ice on the continent of Antarctica. As it melts, it will flow down to the sea and sea levels around the world will rise. If the whole Antarctic ice

icebergs do break off more frequently, we need to understand the process and its effects.

EFFECTS OF LARGE ICEBERGS

Ironically, while we found increased biological activity near large icebergs, the short-term effect of B-15 was that of decreasing the total productivity in the Ross Sea by blocking the annual dispersal of sea ice.[9] A string of large icebergs, most originally part of B-15, formed barriers disrupting the usual flow of sea ice north out of the Ross Sea during the austral

sheet melts, sea levels will rise more than 200 feet, submerging entire island countries and vast regions along the coasts of all continents.

The birth of B-15 had no effect on sea level rise because it was already afloat as part of the Ross Ice Shelf. Its volumetric displacement was already accounted for in the sea level. Was the birth of B-15 a result of global warming? We will never know for sure.

We have not detected the same warming trend in the Ross Sea as has been found in the Antarctic Peninsula region. The cracks in the Ross Ice Shelf that eventually gave birth to B-15 were observed eight years before it calved. Edges of the Ross Ice Shelf break away naturally, but no one had ever seen such a large chunk launched into the sea before. B-15 helps us understand global warming and change; global warming is clearly causing ice shelves to disintegrate in other locations, and scientists predict this trend will continue. If enormous

summer of 2000-2001. The retained sea ice inhibited the penetration of sunlight into the water during those critical 24-hour summer days of explosive plankton growth. Sunlight ignites the photosynthetic process upon which all life in the Antarctic depends. This disruption decreased the overall summertime phytoplankton growth by 50 percent, with effects felt all through the Antarctic food web.

The nature of these icebergs and their biological effects are clearly part of our future. Our journey was a first step in developing critical understanding of iceberg ecology and an attempt to bring a gigantic event of planetary proportions down to human dimensions.

For myself personally, this was a return to a part of the world that I adore for its stark, vast and icy beauty and for the opportunity to study the raw, primal forces of the ocean that play such a vital role in controlling the environmental conditions of our planet.

Series: A Ross seal disappears into a hole in the ice to search for squid, fish and krill.
Photography: C. Olavarría

The Future

We need to understand the connection between our actions and the health of planet Earth.

A year later, the symptoms of frostbite and neuralgia have faded, but the wonder of the expedition never will. The blinding whiteness of icebergs against dark blue Antarctic waters comes back to me at unlikely moments – in the middle of a meeting, sometimes, or while diving in the warm, colorful waters of the South Pacific.

Looking back, our expedition was very successful. Our observations supported research indicating that Antarctica's monstrous bergs, some of which can last for years, are a major factor in the biology of the Ross Sea. As they calve, move and melt, large icebergs play important roles in the operation of the entire global ocean system. That system covers 70 percent of the earth, and what we learn in Antarctic waters gives fundamental clues about the future of our planet.

Will more large icebergs form as a result of global warming? That question was answered one year after I returned from our expedition. On March 20, 2002, I read this headline in *The New York Times*: "Large Ice Shelf in Antarctica Disintegrates at Great Speed." The story reported the breaking apart of a major part of the large Larsen B Ice Shelf on the Antarctic Peninsula. The ice had been there for 12,000 years. Like our exploding iceberg, the ice shelf disintegrated over a very short time period and created thousands of bergs. The area of collapse was 1,300 square miles or 720 billion tons of ice. Over the last five years, the Larsen B Ice Shelf has lost a total of 2,280 square miles of ice, and is now only 40 percent of its previous minimum stable size.

Scientists now understand more about the forces that cause these massive ice shelves to break apart. Through satellite data, pools of melted water were detected on the surface of the Larsen B Ice Shelf prior to its collapse. The pooled water entered cracks in the ice and the weight of the water, like a chisel, forced the cracks to enlarge and eventually caused the disintegration of the ice shelf.

This mechanism of melting, settling and wedging the ice free tells us that the break-up of these ice shelves can happen much faster than previously thought and on larger scales. For example, if water pools began forming on the Ross Ice Shelf, the largest of all ice shelves, it could break apart. If the Ross Ice Shelf breaks apart, ice presently on land could flow more easily and quickly off the continent without anything to slow it down. If the ice flowed into the ocean more quickly, there would be a rise in sea level, as well as an increase in the number of large icebergs adrift in the oceans. Antarctica's massive ice shelves currently act as barriers to this process, but for how long?

What will the future hold? I'm reminded of the saying: "May you live in interesting times." While some see this saying as a blessing, more seem to see it as a curse. Regardless of how you see it, we certainly have interesting times ahead.

We need to understand the connection between our actions and the physical and biological mechanisms of the earth. Although these processes have no doubt occurred numerous times in the geologic past without affecting humans, the present era carries an inherited responsibility to be stewards of our island, planet Earth.

Facing page: Our helicopter pilot Laurie Prouting took time out to visit with an Adélie penguin that shared the ice floe on which he landed.
Photography: P. Turnbull

CHAPTER 7

Global Warming:
What You Can Do

There is now little doubt that the collective actions of people are affecting the earth's climate. It has taken hundreds of years and billions of people to get us to this point. It will take hundreds of years and the cooperation of billions of people to reverse the process and to make the human presence on earth "climate neutral" again. We have the power to do this; the power rests in the policies of governments and the actions of individual people.

THESE ARE SOME WAYS YOU CAN HELP:

Travel More Sustainably:
1. Choose to live close to your workplace to reduce the need to drive.
2. Whenever practical, walk, bike or take public transportation.
3. If you must drive, buy a fuel-efficient, low-pollution car.
4. Think twice before purchasing a second car.

Reduce Household Energy Use:
1. Reduce the environmental costs of heating and hot water with efficient home insulation.
2. Use the sun's energy for heat and electricity.
3. Replace standard light bulbs with energy-efficient compact fluorescent bulbs.
4. Invest in more energy-efficient electronics and household appliances.
5. Buy green power by choosing an electricity supplier that uses renewable energy.
6. Plant trees to absorb more CO_2 and insulate your home against hot sun and cold winds.
7. Reduce, reuse and recycle.

Reduce Your Use of These High Impact Products:
1. Powerboats
2. Gasoline-powered yard equipment
3. Fireplaces and wood stoves
4. Recreational off-road vehicles

Get Involved in Your Community:
1. Support laws and organizations that encourage renewable energy.
2. Tell your elected representatives that you want them to support international treaties that identify global climate change and its solutions.

Facing page: When not clustered at breeding sites near shore, Adélie penguins feed in the ocean. The penguins rely on ice floes for resting and escaping predators.
Photography: P. Turnbull

Endnotes and References

ENDNOTES

[1] Pyne, S.J. *The Ice: A Journey to Antarctica*, 1st Edition. Iowa City: University of Iowa Press, 1986.

[2] Cooley, *K., Moon Facts*. http://home.hiwaay. net/~krcool/Astro/moon/#mf> (17 April 2003).

[3] PBS/NOVA Online, *Antarctic Almanac: Ice Companion site for the NOVA program "Warnings from the Ice."* <http://www.psb.org/wgbh/nova/warnings/almanac.html> (17 April 2003).

[4] Pyne, S.J. *The Ice: A Journey to Antarctica*, 1st Edition. Iowa City: University of Iowa Press, 1986.

[5] Pogo, a witty, political possum created by cartoonist Walt Kelly, was a well-known fixture of American popular culture from the late 1940s to the 1970s.

[6] A nautical mile or "knot" is 6080.4 feet, which is equal to one minute of latitude at the equator. A one-knot current is very strong for a swimmer or diver.

[7] Loeb, V., V. Siegel, O. Holm-Hansen, W. Fraser, W. Trivelpiece and S. Trivelpiece. 1977. Effects of sea-ice extent and krill or salp dominance on the Antartic food web. *Nature* 387:897-900.

[8] The term "Spaceship Earth" was coined by R. Buckminster Fuller, inventor, architect, engineer, mathematician, poet and cosmologist.

[9] Arrigo, K.R. and G. van Dijken. 2000. The impact of the B-15 iceberg on phytoplankton abundance in the Ross Sea, Antarctica, 2000-2001. Paper presented to the American Geophysical Union Conference, May 28 - June 2, 2001. Boston, Massachusetts.

REFERENCES

Antarctic Connection, *Ships of Antarctic Exploration*. <http://www.geocities.com/syendurance/ships.html> (17 April 2003).

Arrigo, K.R. and G. van Dijken. 2000. The impact of the B-15 iceberg on phytoplankton abundance in the Ross Sea, Antarctica, 2000-2001. Paper presented to the American Geophysical Union Conference, May 28 - June 2, 2001. Boston, Massachusetts.

Australian Academy of Science, *The Southern Ocean and Global Climate*. NOVA: Science in the News. <http://www.science.org.au/nova/018/018print.htm> (17 April 2003).

Brower, M. and W. Leon. *The Consumer's Guide to Effective Environmental Choices: Practical Advice from the Union of Concerned Scientists*. New York: Three Rivers Press, 1999.

Carsey, F.D., G.S. Chen, J. Cutts, L. French, R. Kern, A.L. Lane, P. Stolorz and W. Zimmerman. 2000. Exploring Europa's ocean: A challenge for marine technology of this century. *Marine Technology Society Journal* 33(4):5-12.

Cooley, K., *Moon Facts*. <http://home.hiwaay.net/~krcool/Astro/moon/#mf> (17 April 2003).

Dauncey, G. and P. Mazza. *Stormy Weather: 101 Solutions to Global Climate Change*. Gabriola Island: New Society Publishers, 2001.

Diamond, J.M. *Guns, Germs, and Steel: The Fates of Human Societies*. New York: W.W. Norton & Co., 1997.

Earle, S.A. *National Geographic Atlas of the Ocean: The Deep Frontier*. Washington, D.C.: National Geographic, 2001.

El-Sayed, S.Z. *Southern Ocean Ecology: The Biomass Perspective*. Cambridge: Cambridge University Press, 1994.

Fogg, G.E. and D. Smith. *The Explorations of Antarctica: The Last Unspoilt Continent*. London: Cassell Publishers Limited, 1990.

Harrison, P. *Seabirds: An Identification Guide*. Boston: Houghton Mifflin Company, 1983.

Heacox, K. *Shackleton: The Antarctic Challenge*. Washington, D.C.: National Geographic, 1999.

Heinerth, J. Personal Communication. 5 July 2002.

Huntford, R. *The Last Place on Earth: Scott and Amundsen's Race to the South Pole*. New York: The Modern Library, 1999.

International Association of Antarctica Tour Operators (IAATO), *Overview of Tourism September 2000, Agenda Item 13, Submitted by IAATO to the Antarctic Treaty Consultative Meeting XXIV/IP St. Petersberg, Russia, 9-20 July 2001*. <http://www.iaato.org/reports/overview.html> (17 April 2003).

Jackson, P. *Quail Island: A Link with the Past*. Wellington: New Zealand Department of Conservation, 1990.

Jacobs, S.S. and R.F. Weiss, eds. *Ocean, Ice, and Atmosphere: Interactions at the Antarctic Continental Margin*. Washington, D.C.: American Geophysical Union, 1998.

Jeffries, M.O., ed. *Antarctic Sea Ice: Physical Processes, Interaction and Variability*. Washington, D.C.: American Geophysical Union, 1998.

King, P., ed. *Scott's Last Journey*. New York: Harper Collins Publishers, 1999.

Kovaleski, K. *Antarctica: People*. World Facts and Figures: Miscellaneous facts and statistics about every country in the world. <http://worldfactsandfigures.com/countries/special/antarctica.php> (17 April 2003).

Kozloff, E.N. *Invertebrates*. Philadelphia: Saunders College Publishing, 1990.

Land Information New Zealand. *New Zealand Nautical Almanac, 2000 Edition, NZ 204*. Wellington: Land Information New Zealand Publications, 1999.

Landis, J., "Ice of Ages: Where Geometry Meets Geology." *The Antarctic Sun,* 7 January 2001. <http://www.polar.org/antsun/oldissues2000-2001/2001_0107/index.html> (17 April 2003).

Laseron, C.H., F. Hurley and T. Bowden. *Antarctic Eyewitness: Charles F. Faseron's South with Mawson and Frank Hurley's Shackleton's Argonauts*. Sydney: Angus & Robertson, 1999.

Liangbiao, C., A.L. DeVries and C.C. Chang. 1997. Evolution of antifreeze glycoprotein gene from a trypsinogen gene in Antarctic notothenioid fish. *National Academy of Sciences' Proceedings, April 15, 1997* 94(8):3817-3822.

Lizotte, M.P. and K.R. Arrigo, eds. *Antarctic Sea Ice: Biological Processes, Interactions and Variability*. Washington, D.C.: American Geophysical Union, 1998.

Loeb, V., V. Siegel, O. Holm-Hansen, R. Hewitt, W. Fraser, W. Trivelpiece and S. Trivelpiece. 1997. Effects of sea-ice extent and krill or salp dominance on the Antarctic food web. *Nature* 387(6636): 897-900.

MacAyeal, D. and M. Lazzara, *Iceberg Motion*. <http://amrc.ssec.wisc.edu/ amrc/iceberg.html> (17 April 2003).

Marsdon, R. Personal Communication. August 2002.

Mawson, D. *The Home of the Blizzard: A True Story of Antarctic Survival*. New York: St. Martin's Press, 1998.

Maxtone-Graham, J. *Safe Return Doubtful: The Heroic Age of Polar Exploration*. New York: Charles Scribner's Sons, 1988.

Moser, D. Personal Communication. August 2002.

National Snow and Ice Data Center, *Antarctic Icebergs*. http://www.nsidc.org/icebergs/index.html> (17 April 2003).

Neider, C., ed. *Antarctica: Firsthand Accounts of Exploration and Endurance*. New York: Cooper Square Press, 1972.

Norris, B. Personal Communication. August 2002.

PBS/NOVA Online, *Antarctic Almanac: Ice Companion site for the NOVA program "Warnings from the Ice."* <http://www.pbs.org/ wgbh/nova/warnings/almanac.html> (17 April 2003).

Peter's Antarctic Domain, *A Timeline of Early Antarctic Exploration*. <http://www. geocities.com/RainForest/Canopy/8947/ index.html> (17 April 2003).

Preston, D. *A First Rate Tragedy: Robert Falcon Scott and the Race to the South Pole*. Boston: Houghton Mifflin Company, 1997.

Pyne, S.J. *The Ice: A Journey to Antarctica, 1st Edition*. Iowa City: University of Iowa Press, 1986.

Reader's Digest. *Antarctica: Great Stories from the Frozen Continent*. Sydney: Reader's Digest Services Pty Limited, 1985.

Reeves, R.R., B.S. Stewart, P.J. Clapham and J.A. Powell. *National Audubon Society: Guide to Marine Mammals of the World*. New York: Alfred A. Knopf, 2002.

Revkin, A.C. "Large Ice Shelf in Antarctica Disintegrates at Great Speed." New York Times 20 March 2002, late edition - final, section A: page 13.

Rosentrater, L., *Secrets of the Ice: An Antarctic Expedition*. Museum of Science, Boston. <http://www.secretsoftheice.org/ explore/discovery.html> (17 April 2003).

Scambos, T. Personal Communication. 26 July 2002.

Scambos, T.A., C. Hulbe, M. Fahnestock and J. Bohlander. 2000. The link between climate warming and break-up of ice shelves in the Antarctic Peninsula. *Journal of Glaciology* 46:516-530.

Shackleton, E. *The Heart of the Antarctic*. New York: Signet, 2000.

Sierra Club, *Ten things you can do to help cure global warming*. Sierra Club Global Warming and Energy Program. <http://www.sierraclub. org/globalwarming/tenthings.asp> (17 April 2003).

South-Pole.com, *Robert Falcon Scott 1868-1912*. <www.south-pole.com/ p0000089.htm> (17 April 2003).

Stone, G., C. Andrew and S. Earle, eds. 2000. Deep ocean frontiers. *Marine Technology Society Journal* 33(4):3-4.

Sun, L.G. and Z.Q. Xie. 2001. Changes in lead concentration in Antarctic penguin droppings during the last 3,000 years. *Environmental Geology* 40(10):1205-1208.

Thoma, M.P. Golden Gate Bridge Facts. <http://www.thoma.com/ thoma/ggbfacts.html> (17 April 2003).

United States Environmental Protection Agency, *United States Environmental Protection Agency Home Page: Global Warming-Actions*. <http://yosemite.epa.gov/oar/globalwarming. nsf/content/ActionsIndividualMakesaDifference .html> (17 April 2003).

Waller, G., ed. *Sea Life: A Complete Guide to the Marine Environment*. Washington, D.C.: Smithsonian Institute Press, 1996.

Wilford, J.N. "Ages-Old Icecap at North Pole is Now Liquid, Scientists Find." *New York Times* 19 August 2000, late edition - final, section A: page 1.

Wilson, P. 2002. "Too much ice in the drink? A penguin dilemma." Paper presented to the Antarctica New Zealand's Annual Antarctic Conference, 23 and 24 April 2002. Auckland, New Zealand.

Further Reading:

THE "ICE ISLAND" EXPEDITION

Heinerth, J., *Exploring the Greatest Iceberg in History*. <http://www.iceisland.net/team.html> (17 April 2003).

Stone, G.S. Exploring Antarctica's Islands of Ice. *National Geographic Magazine*, December 2001: 36-51.

ANTARCTICA

Baines, J.D. *Antarctica*. Austin: Raintree Steck-Vaughn, 1998.

Debenham, F. *Antarctica: The Story of a Continent*. New York: MacMillan, 1961.

Hargreaves, P. *The Antarctic*. Morristown: Silver Burdet, 1980.

Heacox, K. *Antarctica: The Last Continent: National Geographic Destinations Series*. Washington, D.C.: National Geographic Society, 1998.

Scarf, M. *Antarctica, Exploring the Frozen Continent*. New York: Random House, 1970.

ANTARCTIC HISTORY

Amundsen, R.E., and R. Huntford. *The South Pole: An Account of the Norwegian Antarctic Expedition in the "Fram," 1910-1912*. New York: Cooper Square Press, 2001.

Baughman, T.H. *Pilgrims on the Ice: Robert Falcon Scott's First Antarctic Expedition*. Lincoln: University of Nebraska Press, 1999.

Bickel, L. *Mawson's Will: The Greatest Polar Survival Story Ever Written*. South Royalton: Steerfoth Press, 2000.

Chapman, W. *The Loneliest Continent: The Story of Antarctic Discovery*. Greenwich: New York Graphic Society Publishers, Ltd., 1964.

Flaherty, L. *Roald Amundsen and the Quest for the South Pole, Edition: Library Binding*. Broomall: Chelsea House Pub, 1992.

Fogg, G.E. and D. Smith. *The Exploration of Antarctica: The Last Unspoilt Continent*. London: Sterling Publishing, Co., 1990.

Harrowfield, D.L. *Icy Heritage: The Historical Sites of the Ross Sea Region, Antarctica*. Christchurch: Antarctic Heritage Trust, 2002.

Heacox, K. and A. Shackleton. *Shackleton: The Antarctic Challenge*. Washington, D.C.: National Geographic Society, 1999.

Headland, R.K. *Chronological List of Antarctic Expeditions and Related Historical Events (Studies in Polar Research)*. Cambridge: Cambridge University Press, 1990.

Huntford, R. *Shackleton*. New York: Carroll & Graf, 1998.

Huxley, E.J.G. *Scott of the Antarctic*. London: Weidenfeld and Nicholson, 1990.

Jacka, F. *Mawson's Antarctic Diaries*. London: Routledge, 1990.

Morrell, M., S. Capparell, and A. Shackleton. *Shackleton's Way: Leadership Lessons from the Great Antarctic Explorer*. New York: Viking Press, 2001.

Ralling, C. *Shackleton: His Antarctic Writings Selected and Introduced by Christopher Ralling*. London: British Broadcasting Corporation, 1983.

Ross, M.J. *Polar Pioneers: John Ross and James Clark Ross*. Montreal: McGill-Queens University Press, 1994.

Scott, L. and H.G. Ponting. *The Great White South: Traveling with Robert F. Scott's Doomed South Pole Expedition*. New York: Cooper Square Press, 2002.

Scott, R.F. *Scott's Last Expedition: The Journals*. New York: Carroll & Graf Publishers, Inc., 1996.

Scott, R.F. *The Voyage of Discovery: Scott's First Antarctic Expedition*. New York: Cooper Square Press, 2001.

Stewart, J. *Antarctica: An Encyclopedia: M-Z Chronology, Expeditions, Bibliography*. Jefferson: McFarland & Company, 1990.

Stuster, J. *Bold Endeavors: Lessons from Polar and Space Exploration*. Annapolis: Naval Institute Press, 1996.

ANTARCTIC RESEARCH

Fogg, G.E. and M. Thatcher. *A History of Antarctic Science (Studies in Polar Research)*. Cambridge: Cambridge University Press, 1992.

Kennett, J.P. and D. Warnke, eds. *The Antarctic Paleoenvironment: A Perspective on Global Change*. Washington, D.C.: American Geophysical Union, 1992.

MacAyeal, D. and M. Lazzara, *Iceberg Motion*. <http://amrc.ssec.wisc.edu/amrc/iceberg.html> (17 April 2003).

National Aeronautics and Space Administration, *West Antarctic Sheet Initiative*. http://igloo.gsfc.nasa.gov/wais/> (17 April 2003).

National Snow and Ice Data Center, *National Snow and Ice Data Center Home Page*. <http://nsidc.org/index.html> (17 April 2003).

Rosentrater, L. *Secrets of the Ice: An Antarctic Expedition*. Museum of Science, Boston. <http://www.secretsoftheice.org/index.html> (17 April 2003).

Scientific Committee on Antarctic Research (SCAR), *Scientific Committee on Antarctic Research (SCAR) Home Page*. <http://www.scar.org/> (17 April 2003).

ANTARCTIC NEWS

70 South. <http://www.70south.com/home> (17 April 2003).

The Antarctic Sun.<http://www.rpsc.raytheon.com/antsun/index.htm> (17 April 2003).

INTERNATIONAL POLICY

The Antarctica Project. <http://www.asoc.org/> (17 April 2003).

National Science Foundation, *United States of America Antarctic Treaty Information Exchange.* <http://www.nsf.gov/od/opp/ antarct/treaty/> (17 April 2003).

TerraQuest, *Virtual Antarctica.* <http://www.terraquest.com/va/bridge/bridge.html> (17 April 2003).

ANTARCTIC ECOLOGY

Adrian, M. *Wildlife in the Antarctic.* New York: J. Messner, 1978.

Gorman, J. *Ocean Enough and Time: Discovering the Waters around Antarctica.* New York: HarperCollins, 1995.

Soper, T. *Antarctica: A Guide to the Wildlife.* Bucks: Bradt Publishings, 2001.

Stonehouse, B. *Animals of the Antarctic: The Ecology of the Far South.* New York: Holt, Rinehart Winston, 1972.

Van Mieghem, J. and P. Van Oye, eds. *Biogeography and Ecology in Antarctica.* The Hague: W. Junk, 1965.

ANTARCTIC CONSERVATION

Antarctic and Southern Ocean Coalition, *Antarctic and Southern Ocean Coalition Home Page.* <http://www.asoc.org/> (17 April 2003).

Gland: International Union for Conservation of Nature and Natural Resources. *A Strategy for Antarctic Conservation,* 1991.

Timberlake, L., ed. *Conservation and Development of Antarctic Ecosystems.* Gland: International Union for Conservation of Nature and Natural Resources, 1984.

WHAT YOU CAN DO

Brower, M. and W. Leon. *The Consumer's Guide to Effective Environmental Choices: Practical Advice from the Union of Concerned Scientists.* New York: Three Rivers Press, 1999.

Dauncey, G. and P. Mazza. *Stormy Weather: 101 Solutions to Global Climate Change.* Gabriola Island: New Society Publishers, 2001.

United States Environmental Protection Agency, *Global Warming-Actions.* http://yosemite.epa.gov/oar/globalwarming.nsf/content/Actions IndividualMakeaDifference.html> (17 April 2003).

Acknowledgments

The Bermuda Underwater Exploration Institute, the National Geographic Society, the New England Aquarium, Karst Productions and Kurtis Productions provided essential support for this expedition and book. I appreciate the cooperation of the New Zealand Ministry of Foreign Affairs and the New Zealand Antarctic Programme for permits and permission to conduct this research expedition under protocols of the Antarctic Treaty.

I thank New England Aquarium President Edmund Toomey, Board Chair Susan Solomont, Trustees and Overseers for their positive energetic leadership of an organization that encourages projects like *Ice Island*. I thank my shipmates at the Bermuda Underwater Exploration Institute for sharing a common vision of exploration and stewardship for the world's oceans including Teddy Tucker, Edna Tucker, Wendy Tucker, Peter Benchley and Michael Collier. It was a pleasure working with the *National Geographic Magazine* on this project and I thank Bill Allen, Emory Kristof and Kathy Moran.

I thank Wes Skiles, my co-expedition leader, for his brilliant photography and companionship on this amazing adventure. Thanks to my good friend Porter Turnbull, my dive buddy in the ice and our ship's medical officer. I could not have done this without you, Porter. Thank you, Carlos Olavarría, for your excellent research work under extraordinary conditions. I thank Jill and Paul Heinerth, two of the hardest working and best divers I know. I thank the crew and captain of *Braveheart* for delivering us safely to and from the Ross Sea.

Thanks to Cynthia Nichols for her enthusiasm and help with the hundreds of details in preparation for this expedition and this book. Thanks to my colleague Heather Tausig for her tremendous management and leadership on all other projects at the New England Aquarium that would otherwise have suffered while I was out to sea for two months. I thank my good friends and colleagues at the New England Aquarium, the National Geographic Society and the Bermuda Underwater Exploration Institute for support, help and consultation: Joanna Allen, John Anderson, Steve Bailey, Al Barker, John Constable, Michael Connor, Walter Flaherty, Liz Gorham, Noela Haycock, Ellen Hurley, Peter Johnson, Scott Kraus, Sandy Lane, Robert Lippincott, Ken Mallory, Debie Meck-Maher, Karen Mize, Keith Moorehead, Susan Nelson, José Luis San Miguel, John Prescott, Sandra Prescott, Cristina Santiestevan, Tedd Saunders, Jerry Schubel, Crystal Shultz, Caroly Shumway, Billy Spitzer, Carmine Tocci, Charlotte Wilson, Sheldon Wolf and Joe Zani. I thank George Matsumoto, Bruce Robison and Larry Madin for help identifying the jellyfish and siphonophores and to Paul Dayton for his help with the analysis of benthic marine life. I thank Karsten Hartel and Joe Eastman for identifying icefish specimens, and Harlan Dean for identifying polychaetes.

Thank you, Bill and Peggy Hamner, for showing me the way to Antarctica and for teaching me how to dive there in 1987. Thanks to Meri Laesk for essential communications help. Thanks to Scott Baker, Mildred Crowell, Mike Donoghue, Alan Dynner, Alistair Hutt, Steve Katona, Susan Lerner, Akiko Shiraki and John Viele. Thanks to John Splettstoesser and James McCarthy for comments on the manuscript. Thanks to Doug MacAyeal, Dave Moser, Baden Norris and Ted Scambos for historical and background information. For keeping me warm above and below water, thanks to SIERRA DESIGNS® and OCEANIC®.

Special praise and thanks to the team that worked endlessly and diligently on the production of this book. Jennifer Goebel is a superb editor and helped organize the flow of my story, helped me part with boring sections, and dedicated many hours to improving the manuscript. Catherine LeBlanc and Jonathan Place are excellent, talented and extremely hardworking designers: they gave the book its look and feel. I am grateful for Catherine Fox's passion and work on this project, from helping design illustrations, to researching facts and figures, and helping with analysis of CTD data. Thanks to Lisa Spalding for her dedication and flawless work on all aspects of her work at the Aquarium and especially for essential help with the production of this book. I thank Wes Skiles, Porter Turnbull, Carlos Olavarría, Jill and Paul Heinerth, and the Canterbury Museum, Christchurch, New Zealand, for use of photographs. Additional illustrations were provided by the National Geographic Society, the National Ice Center, the Naval Ice Center, and the NASA/Goddard Space Flight Center.

Finally, loving thanks to my wife, Austen Yoshinaga, for encouraging me to go on this adventure and for her constant support in all aspects of my life. And to my family: Connie, Charles, Greta, Andy, Jonathan, Maddy, Juan, Roger and Alma for many years of support and understanding.

Photography: G. Stone

ICE ISLAND
Expedition to Antarctica's Largest Iceberg
By Gregory S. Stone

Copyright © 2003 by the New England Aquarium

The New England Aquarium's mission is to present, promote and protect the world of water.

First published in 2003 by New England Aquarium Press in association with the
Bermuda Underwater Exploration Institute

Distributed by Bunker Hill Publishing, Inc.
26 Adam Street, Charlestown, Massachusetts 02179 USA
6 The Colonnade, Rye Road, Hawkhurst, Kent TN184ES UK
www.bunkerhillpublishing.com

Text copyright © 2003 by Gregory S. Stone
Photographs copyright © 2003: *Wes Skiles, Porter Turnbull, Gregory S. Stone, Carlos Olivarría,
Jill Heinerth, Paul Heinerth, Canterbury Museum, Mary Jane Adams, Catherine Fox*
Illustrations copyright © 2003: *National Geographic Magazine, National Geographic Maps,
New England Aquarium, W.W. Norton & Company, Inc.*

ISBN: 1-59373-017-9

Printed by
Henry N. Sawyer Company, Inc.
586 Rutherford Avenue, Charlestown, Massachusetts 02129

Cover photo: *Wes Skiles*
Book design: *Catherine LeBlanc and Jonathan Place*
Project editor: *Jennifer Goebel*
Research assistants: *Catherine Fox and Lisa Spalding*
Production support: *Cynthia Nichols and Austen Yoshinaga*
Photographs: *Wes Skiles, Gregory S. Stone, Porter Turnbull, Canterbury Museum (New Zealand),
Carlos Olivarría, Jill Heinerth, Paul Heinerth, Mary Jane Adams, Catherine Fox*
Illustrations: *Barbara Harmon, National Geographic Magazine, National Geographic Maps,
W.W. Norton & Company, Inc.*

Printed in the United States of America

10 9 8 7 6 5 4 3 2 1